INNER UNIVERSE

INNER UNIVERSE

Fundamental Science,
Phenomenology &
Fields of Awareness

Peter Wilberg

New Yoga Publications

First published by New Yoga Publications in 2012
www.thenewyoga.org

Printed by CreateSpace.com
ISBN 10 1489597441
ISBN 13 9781489597441

Cover design by LucidArt.co.uk

"However vast outer space may be, yet with all its sidereal distances it hardly bears comparison with the dimensions, *with the depth dimension of our inner being*, which does not even need the spaciousness of the universe to be within itself almost unfathomable... To me it seems more and more as though our customary consciousness lives on the tip of a pyramid whose base within us (and in a certain way beneath us) widens out so fully that the farther we find ourselves able to descend into it, the more generally we appear to be merged into those things that, independent of time and space, are given in our earthly, in the widest sense, worldly existence."

Rainer Maria Rilke

Contents

Fundamental Science or Fundamentalism?

> "Philosophy is Primordial Science".
>
> Martin Heidegger

In an age in which rigid religious fundamentalisms conflict or coexist uneasily with no less rigid scientific fundamentalisms – including both bio-genetic and quantum-physical reductionism – is it anymore possible to articulate a fundamentally new understanding of the universe?

In an age in which post-modernism has relativised the words and works of the world's greatest philosophical thinkers, from Heraclitus to Heidegger – and our universities have reduced thinking itself to merely aping and applying the unquestioned jargons of different specialist sciences and academic fields – can one any more claim to propound a fundamentally new philosophy or a fundamentally new understanding of 'science' as such?

These questions are important since today science and religion stand at a threshold – both having failed to provide a philosophically grounded and consistent account of the fundamental nature of reality. Religious fundamentalisms substitute deep understandings of the fundamental nature of reality with symbols of this reality. Meanwhile, modern science – and physics in particular – has retreated into its own form of quasi-religious fundamentalism, taking its own purely quantitative mathematical symbols and conceptual constructs as more fundamentally 'real' than the qualitative, tangibly experienced 'empirical' phenomena they are used to 'explain'.

What alone can truly be called 'fundamental' science has nothing in common with either religious or scientific fundamental*isms*. Its foundation lies instead in a fundamental realm of truth ignored or misconceived in both atheistic science and theistic religions. This is the realm of non-extensional or 'intensional' reality that I call the 'Inner Universe' – a realm of pure *potentialities* of existence or being within which all extensional or 'space-time' universes first open up.

The Inner Universe is not a purely quantitative universe of numbers denoting so-called energetic states or 'quanta' but is composed instead of tangible elements of qualitative experience or 'qualia'.[1] It is made up not of matter, energetic quanta or energy fields but of fields, field-patterns and field dynamics of *awareness*. For on the most fundamental level, awareness alone and as such can be understood as *the* most 'fundamental' reality of all – being irreducible, in principle, to a mere 'property' or 'product' of any phenomena which we only know of *through* it (whether in the form of matter or energy, a God or gods, our bodies or our brains).

Fundamental Science is a 'field-phenomenological science' of awareness. What this means is that it acknowledges the 'non-local' or 'field character' of awareness. Together with this goes the recognition that (1) *all* experienced phenomena emerge from and within fields of awareness and constitute manifestations of that field (2) that *no* field of awareness can be reduced to or explained by any phenomena emerging from it and (3) that since *all* phenomena emerge from fields of awareness, *no* phenomena can be explained as simply 'caused' by other phenomena emerging from the same field.

Fundamental Science, as field-phenomenological science, also constitutes a 'unified field theory of awareness' – based on the

---

[1] see my more recent book entitled *The Qualia Revolution – from Quantum Physics to Cosmic Qualia Science.*

recognition of a 'universal awareness field' of which all phenomena and all things, all beings, all worlds and all dimensions of reality are an aware and embodied expression.

In this way it also offers a more fundamental understanding of both the physical and life sciences than they themselves can provide. The physics beyond physics or 'meta-physics' of Fundamental Science also provides the theoretical and practical foundations for a medicine beyond medicine, a 'meta-medicine' that challenges the reduction of the human being and human consciousness to the human body and brain. In doing so it offers a radical alternative to the ever-increasing use of biotechnology, a technology whose sole aim, besides corporate profit, is the annihilation of all bodily and behavioural expressions of individual and social dis-ease.

The result of over thirty years of phenomenological research, Fundamental Science, newly understood as Field-Phenomenological Science, is not just another 'New Age' miscegenation of the latest development in the sciences with poorly understood or superficially rehashed metaphysical or 'spiritual' traditions of the past. Instead it is a radical philosophical rethinking of the fundamental nature of 'science', 'scientific method', 'scientific research' – and 'scientific' knowledge and truth.

Drawing, among other sources, on the Seth books of Jane Roberts, the Spiritual Science of Rudolf Steiner, the 'Dialectic Phenomenology' of Michael Kosok, Rupert Sheldrake's theory of Morphic Resonance and Eugene Gendlin's philosophy of directly felt meaning or "felt sense". In addition, it offers new methods of field-phenomenological *research* - which uses no instrument but the researchers' own 'intensional body' or 'inwardly felt body' – itself a body of feeling awareness.

With it, new forms of joint meditational and experiential exploration of the 'inner universe' can and have been conducted and *inter-subjectively* validated, i.e. their 'results' confirmed and duplicated not through quantitative measurements but rather through the qualitative evidence offered by the shared experiential findings of different researchers.

Fundamental Science and the Logos

Religious accounts of fundamental 'truth' or 'reality' have always taken speech and language – the Word – as the highest expression of the inner order of the universe, what the Greeks called its *logos*. *Logos* means 'word' or 'account'. Our current concept of 'science' on the other hand understands the *logos* not as an account of an inner or 'implicate' order based on language but rather as one based on numbers – on counting and on formal mathematical 'accounting'.

The Greek verb *legein*, from which both the ancient concept of the *logos* and modern scientific terms such as 'logic', 'biology', psychology etc. derive, contains within itself the seeds of division between science and religion, bearing as it does the double meaning of (a) to understand or 'gather' something and to provide a verbal account of it and (b) to gather things (the harvesting of fruit for example) and to count them. What fundamentalist religion and science both share is the belief in a single pre-given order of things, natural or divine, a single pre-ordered reality. Both science and religion see nature and man as obeying physical or ethical 'laws' which themselves are nothing more nor less than human representations of reality.

Where they both fall short is in providing an account, not of an actual pre-ordered reality but of order as such and what precedes it – of *pre*-order as such. Pre-order is identified with *chaos*, another Greek concept. Science is only just beginning to understand that chaos is not merely random disorder but possesses an intrinsic order of a different sort. But the mathematics of 'chaos theory' is a far cry from the understanding that all ordered systems are the expression of different potential orders or patterns of manifest reality – and that pre-order, far from being a uniform, undifferentiated or formless state of matter

13

or consciousness, is a highly differentiated field of *potential* orders or patterns of manifestation. In other words, any actual or manifest order of things, any actual or manifest phenomena or universe is but one self-manifestation of this primordial field of potential ordered realities or universes.

Fundamentalisms of all sorts seek a foundation for the known universe in terms that are by no means 'fundamental', but derive instead from unquestioned scientific and religious myths – from misconceived interpretations of Fundamental Reality. What I call Fundamental Science has nothing in common with religious or scientific fundamentalisms. Its foundation lies in a dimension of reality misconceived in science and religion, therapy and theology. This is the realm of 'non-extensional' or 'intensional' reality that I call the 'Inner Universe'. This inner universe is made up neither of energetic fields nor their expression in any observed or experienced phenomena we are aware of. Instead it is made up of fields and field-patterns of awareness as such.

Unlike the physical sciences, what is known as 'phenomenological' science has always recognized that subjectivity or awareness *as such* is the pre-condition for the experience of any phenomenon or universe whatsoever – being irreducible, in principle to any thing or being we are aware *of.* What it has not fully explored is the field character and field-dynamics of subjectivity as such. Instead, the objective world is seen as a world of pre-given objects and subjectivity is seen as the property of a pre-given subject, human or divine – thus reducing God to one being or subject among others.

Fundamental Science is 'field-phenomenological science', based on a 'field-dynamic phenomenology'. Its basis is the recognition that fields and field-patterns of awareness are the precondition, not just for our own human experience of *physical* phenomena but for the very emergence [Greek *phusis*] of those phenomena in nature itself. All actual energetic field-patterns have their ultimate source in potential field-patterns. These *potential* field-patterns, by their very nature, form no part of any *actual* extensional universe but have an intensional reality only in awareness as such.

In Husserl's 'Transcendental Phenomenology', awareness is always 'intentional', an awareness of something. On the other hand, Fundamental Science as Field-Phenomenology does not identify awareness purely with consciousness of any actual phenomena but also and more primordially with a primordial field of unmanifest *potentialities* of awareness.

A new Fundamental Science and Theology are united in the understanding that this awareness of potentiality can re-link us not only with our own innermost potentials and those of others, but with the 'Power of God' – that infinite and inexhaustible field of potentiality that constitutes the aware inwardness of all things.

Physical science has its holy Einsteinian trinity – Mass, Energy and Light. But the physical universe and all physical phenomena – including light itself – are only visible or measurable *in the light* of an awareness of them - or of the very instruments with which we measure them. The missing third element in the dualistic science of matter and energy is not the quantity [c] – the speed of light – but consciousness or awareness [a].

In a fundamentally unaware universe of matter and energy no fundamental reality can be attributed to *pre*-order, consisting as it does of *potential* patterns or events rather than manifest or actual ones. Why? Because *potential* patterns, structures and forms by definition have no 'objective' reality as *actual* material or energetic patterns. They have reality only in so far as there is a (subjective) *awareness* of them – only in fields of awareness, and as potential field patterns and qualities *of* awareness.

Fundamental Science therefore, is founded on the revolutionary proposition that awareness is the very inwardness of energy, just as matter is its outwardness. It distinguishes however, between awareness in the form of our own localized human 'consciousness' of the universe, and awareness in its 'non-local' or *field* character.

The religious belief that the universe is a product of a God or gods, of beings with their own consciousness, fails to recognize, that any given being or consciousness is itself the actualisation of one particular ordering or pattern of awareness, and as such is itself but one self-manifestation of a primordial field of awareness consisting of a limitless number of potential patterns and consciousnesses – potential beings. Even monotheistic accounts of reality have always fallen into contradiction by representing God both as an infinite and absolute being and as one being or consciousness among others – whether other gods, mortals, or hierarchies of semi-divine spiritual beings.

The new Fundamental Science is therefore also a Fundamental Theology – in contrast to *fundamentalist* religions. It transcends the self-contradiction of religion by recognizing instead that any true 'God' is not and cannot be one god, being, or consciousness among others but must instead be understood as a primordial field or 'ground-state' of awareness that is the source of all potential or possible gods, beings, or consciousnesses. Conversely, any actual being is both an independent consciousness or 'self' imbued with its own unique pattern of awareness, and, on the other hand the divine self-manifestation of that inexhaustible field of potential patterns of awareness – that we call God. This field is the source not only of a limitless number of actual consciousnesses or beings but of a limitless number of worlds – any such world being nothing but a patterned field of conscious experience, already ordered and shaped by an underlying field pattern of awareness. What we call God is not one being among others that happens to have or possess consciousness. Instead God *is* consciousness – not a consciousness that is 'yours' or 'mine' – the private property of a being or beings – but one that is the very essence of the divine – a field awareness that is the source of all beings.

The Sources of Fundamental Science

1. Eugene Gendlin's concept of directly experienced meaning or felt sense, and its significance for a direct feeling cognition of intensional reality.

2. Rupert Sheldrake's concept of 'morphic resonance' – a fundamental dynamic governing the relation of intensional and extensional reality.

3. Rudolf Steiner's spiritual-scientific understanding of intensional reality as the inner source of both human nature and the natural world itself.

4. Michael Kosok's 'dialectic phenomenology' – a scientific account of the non-local or field character of subjectivity and its field-dynamics.

5. Karl Marx's vision of a unitary and Fundamental Science that would constitute both a "natural science of man" and a "human science of nature".

6. Seth's direct insights into the nature of intensional reality and the inner universe, as presented in the books of Jane Roberts.

7. Martin Buber's understanding of the I-Thou relation as a direct intensional relation between two beings in contrast to an extensional 'I-It' relation.

8. Martin Heidegger's critique of the false metaphysical assumptions underlying the modern scientific-technological world view.

9. Jakob Von Uexküll's understanding of living organisms as aware beings shaping their own unique sensory environment or 'Umwelt'.

10. Meister Eckhart's theology of God's unbounded inwardness or intensional reality. "…what flows out remains within."

11. Heraclitus's 'Cosmic Fragments' – sayings of the first fundamental scientist.

Fundamental Science and 'Therapy'

Therapies, understood in the broadest sense to include both psychotherapy and somatic medicine, have today become forms of religious science and scientific religion, both of which confuse inward and outward, intensional and extensional, potential and actual dimensions of reality. 'Scientific' medicine in particular shows signs of this, with biological reductionism as its basic credo, the corporate bio-tech labs as its temples, and the human genome as its holy book. But just as a book is the outward or extensional manifestation of an invisible, intensional space of meaning – one, which cannot be reduced to patterns of ink marks on a page – so also is the molecular text of the human body. Combinations of genes, like combinations of letters in words are signifiers of an intensional reality, which consists of a field of potential patterns of significance that is irreducible to any actual pattern. Somatic medicine reduces this realm of potential significance – the patient's felt dis-ease and its felt meaning – to a set of diagnostic signs and genetically programmed disease patterns. In no less zealous a fashion, many forms of psychotherapy seek to reduce the client's felt dis-ease and the felt meaning of this disease to a set of outward signifiers of another sort, confusing potential patterns of significance with their actual expression as cognitive or behavioural, verbal or emotional patterns.

Just as theology can easily replace the direct felt sense of God's reality and identify it with symbols of it, so in therapy can a direct felt sense of a person's dis-ease be identified with its outward signs. To recognize through some outward sign that a person is in a certain type of pain or distress, or that they are experiencing a specific emotion such as sadness or guilt, and to feel empathy for them is one thing. To feel the specific quality of this person's pain or distress, this person's sadness or guilt, and to resonate with its unique tonality, is quite another. To be aware of an emotional sign or signifier is not the same thing as to attune to it as a specific tonality of awareness. To register and respond to a sign is not the same thing as to resonate with the specific feeling tone that it gives form to.

The substitution of the symbol or formal signifier for directly felt meaning or significance — for felt sense and a felt resonance with God or with other human beings — is itself the mark of fundamentalisms of all sorts. What Marx called the fetishism of the commodity is just one form of the post-modern fetishism of the signifier, shared by both somatic medicine and psychotherapy in the name of 'health'. What they both find a substitute for is the ethical renewal of the health of human relations as such, one that can only come about through a rediscovery of the intensional dimensions of our relatedness to one another, to nature and to God — a direct felt resonance with the inwardness of both people and things that is a field-resonance in the most fundamental scientific sense. True health is intimately connected with the health of social relations as a medium of individual value fulfilment. Fundamental values are nothing actual — they are not reducible to brand values or corporate values, commodity or market values, cultural or ethnic values, or the purely quantitative mathematical 'values' that form the basis of our current paradigm of 'science'. Fundamental values are the fundamental facts of intensional reality, consisting not of actual but potential qualities of awareness embodied not only in human relations but in all the qualitative dimensions of nature itself.

When we feel moved by a sunset we are not simply projecting human emotions onto nature. We are responding to the felt tonalities of awareness expressed in its colours. Intensional reality consists of landscapes of feeling tone — shaped tonalities and textures of awareness that are the inner source not only of specifically human emotions but of natural phenomena as such. This basic proposition of Fundamental Science is that intensional fields of awareness are the condition, not only for our experience of the natural phenomena, but for their very emergence as phenomena. But understanding the fundamental scientific truth of this proposition means overcoming the prejudice that conscious awareness as such is the exclusive property of human beings or animals – a mysterious end-product of cosmic evolution. This is the prejudice overcome by field-dynamic phenomenology, which recognizes that awareness is not merely the property of a localized bodily subject or ego, but has a non-local or field character. Just as matter in all its forms has its source in basic wavelengths and wave envelopes of energy – in so-called quanta – so consciousness in all its forms has its source in basic qualitative wavelengths and envelopes of awareness – in qualia.

Fundamental Science and Theosophy

Drawing on Eastern spiritual and scientific traditions, in particular Vedic philosophy and science, theosophists have long argued that there are basic units or 'atoms' of awareness as well as basic units of matter or energy. In doing so they provided a bridge between idealistic and materialistic philosophies, between theology and theoretical physics. This was a bridge that was otherwise lacking in Western thought until the philosophical implications of quantum theory began to sink in — bringing into question as it did our understanding of what matter as such essentially is. Theosophical science has nevertheless been studiously ignored by both scientists and academic philosophers of science, as well as by clerics and theologians of all faiths — regarded as eccentric or fanciful speculation not worthy of any serious intellectual attention. The principal threat that theosophy posed to both science and religion lay in transgressing the politely respected boundaries between scientific 'knowledge' and simple religious 'faith'. It did this by acknowledging an inner dimension of Creation that could itself be the object of direct cognition and scientific investigation — an inner universe consisting of countless non-extensional fields or planes of awareness, an inner universe that linked the physical reality-field in which we dwell as human beings to the withinness or intensional reality of God. Theosophy challenged what Heidegger called the 'double accounting' of the scientist as a human being — that sort of schizophrenia that allows a human being to appreciate fine art or even follow a faith whilst at the same time, as a scientist being forced to stick to the official line that both our perception of a work of art and the very concept of a God are products of chemical processes occurring in the brain.

The limits of theosophy lay in its lack of a properly developed phenomenological basis. Theosophy failed to recognize, as phenomenology did, that science rested not so much on sense perceptions or empirical facts as on theoretical models and their verbal signifiers — on concepts such as 'matter', 'energy', 'particle' and 'waves' etc. These concepts were understood in a naïve way as names for pre-existing things. And just as conventional science saw the outer universe as a world of pre-given material entities and physical forces, so did theosophical or spiritual science see the inner universe as a neatly ordered world of pre-given spiritual beings and non-physical forces. Theosophy, however, was more explicit than science in taking signifiers as its starting point — not just verbal concepts or mathematical signs but esoteric symbols inherited from archaic spiritual traditions. Just as the empirical scientist's research was unknowingly framed by unquestioned theoretical models and metaphors, so was the theosophist's own direct 'psychical' or 'clairvoyant' cognition of the inner universe generally framed by these occult symbols and the unquestioned spiritual doctrines that were their source. Alternatively, theosophical 'knowledge' was expressed in terms borrowed from the science of the day, but in a way that did not question those terms any more deeply than science itself. Fundamental Science is a fundamental reinterpretation of theosophical knowledge, the modes of cognition and modes of research on which it is based, but one which does not take as its unquestioned foundation either modern scientific concepts or terms borrowed from ancient spiritual doctrines. In this respect it is inspired by the type of theosophy propounded and practiced by Rudolf Steiner under the name 'anthroposophy' — one which recognized that scientific knowledge of the inner universe needed as its foundation clearly defined concepts and modes of cognition.

Fundamental Concepts

Any truly Fundamental Science must address fundamental questions. This includes questioning the fundamental nature of 'science' and scientific knowledge themselves. Most current attempts to provide a new account of the fundamental nature of reality tend to draw on concepts deriving both from specific fields of modern (Western) science on the one hand and on specific (usually Eastern) religious or mystical traditions on the other. In doing so, they commit a fundamental error. For Fundamental Science cannot, by its very nature, be a grand unifying 'synthesis' or 'integration' of concepts deriving from specific sciences, specific religions or spiritual traditions, or specific disciplines such as logic and mathematics, linguistics and semiotics, theology and philosophy. Instead its foundation must lie in Fundamental Concepts. Fundamental Concepts are concepts, which, though they may derive from one specific science, one religion or one discipline are in some way comprehensive — fundamental to all sciences, religions and disciplines. As a result they are concepts, which cannot, by their very nature, be understood in the terms of one science, religion or discipline alone. The fact that a Fundamental Concept derives from a specific science such as physics, a specific spiritual tradition such as Buddhism, or a specific discipline such as psychoanalysis does not mean that this science, tradition or discipline is itself somehow fundamental to all others. The reason why I counterpose Fundamental Science to 'physical science' is that in our current concept of science as such, physics is still regarded as a science that is in some way more 'fundamental' than other sciences, and therefore fundamental to them. This is understandable, because many of the basic concepts of physics such as 'field', 'resonance' or 'energy' are indeed Fundamental Concepts. The problem is that they are not understood as such, but instead seen as purely physical concepts, to be understood within the terms of physics alone. The impossibility of doing so is the reason why physics as a science is in such a quandary at the moment, unable to give an adequate account of the fundamental meaning of its own basic concepts, unable to say what 'matter', 'energy', 'electromagnetism', 'gravitation' etc. essentially are. Instead they are defined in terms of one another, as abstract mathematical variables with no intrinsic meaning outside their own mathematical relationship.

Any specific science, whose basic concepts were understood as Fundamental Concepts, would indeed be fundamental to the understanding of all other sciences and have direct relevance to understanding them in a deeper, more fundamental way. A truly Fundamental Physics, for example would be fundamental not just to chemistry and biology, but also to philosophy and psychology, sociology and economics, semiotics and linguistics etc., just as any of these sciences would, as Fundamental Sciences, be fundamental both to each other and to physics itself. Liver, kidneys, lungs, heart and brain all being fundamental to the life of our bodies, the latter cannot be reduced to the functioning of any of these organs, nor can any one of them be regarded as fundamental – the organ of life. The very nature of Fundamental Science lies in the fact that different Fundamental Sciences, like different vital organs of the same body, are all fundamental to one another – with none being more fundamental than all the others. A Fundamental Psychology, or a Fundamental Linguistics or Biology, is at the same time a Fundamental Physics. A Fundamental Philosophy or Theology is at the same time Fundamental Science – and vice versa. That is because at the foundation of all the sciences are Fundamental Concepts that are exclusive to any given science and cannot be understood within the terms of that science alone.

The concept of 'resonance' deriving from physics has a fundamental relevance for psychology, biology and linguistics. Because of this, it has a fundamental meaning that cannot be understood in terms of physics alone, but requires reinterpretation in the context of each of these other sciences and disciplines to be fully understood as a comprehensive or Fundamental Concept. An example of a Fundamental Concept is Rupert Sheldrake's concept of 'morphic resonance' and 'morphic fields'. The concept of 'fields', like that of 'resonance' derives from physics. Sheldrake applies it to biology, using it to account for the genesis of biological form (*morphe*) in a way that biological science itself has so far failed to do, and is essentially incapable of doing. In this paper I show how language and speech, rather than biology, are the most obvious illustrations of the nature and working of morphic fields and morphic resonance as Sheldrake explains them. In doing so however, I not only present his concepts as something fundamental to linguistics, I also re-present them as concepts that can only be fully understood through linguistics as well as through biology and physics. The result is to show that the concepts of 'morphic fields' and 'morphic resonance' are not fundamentally biological, physical or linguistic but that, whatever their origin, they are quite simply Fundamental Concepts pertaining to a Fundamental Science.

Fundamental Science and Technology

Modern physical science is the direct inheritor of a religious world view which saw the universe as an already formed or created system – a closed system. The first law of thermodynamics, which declares that energy cannot be created or destroyed, but only transformed, denies the possibility of a universe that is constantly being created, constantly in-creation and in-formation. In our digital era, information itself is seen as broken down into units – into bits or bytes that only require storage or processing. In reality of course, binary units of information are the expression of a formative activity that is not only processing existing patterns of information but generates new potential patterns in the very process of doing so.

Modern physical science not only inherited a world view from religion. It also brought with it a new type of miracle – the 'miracles' of modern technology. The common belief that this technology is a product of science is, as Heidegger emphasised, a fundamental misconception. Physical science is technological in its very essence, not by virtue of lending itself to the design and production of machines but by viewing the world as a machine i.e. as a closed system. The early scientific picture of the cosmos as a huge clockwork mechanism was an example of the technological essence of closed-system science. Of course, machines have their inputs and outputs, but these are conceived scientifically in an essentially technological way – as part of ever larger, more inclusive systems. That which remains unaccounted for within any closed system is not seen as something outside or external to the system. Instead it is reduced, as Heidegger pointed out, to a "standing reserve" of industrial parts and labour, raw materials or human inputs to the system. The aeroplane on the tarmac and all of its parts exist as such a standing reserve for the system of operations that constitute the airline industry. Similarly, patients and their complaints constitute a standing reserve of inputs for the health industry, or a standing reserve of consumers for its outputs.

Fundamental Science is in accord with New Energy Science in rejecting the closed-system paradigm, and envisaging not only a new open-system science but new open-system technologies and energy sources. It is based on the understanding that there can be no such thing as a closed system or bounded universe, any such boundary implying a reality beyond it. In particular however, Fundamental Science introduces the fundamental and revolutionary concept of a universe that is inwardly rather than outwardly unbounded, and capable of infinite inward expansion. This is a universe whose extensional dimensions open up within intensional spaces and fields that occupy no bounded extensional space whatsoever.

Fundamental Research

Fundamental Science is a fundamental rethinking of the very nature of science and scientific investigation, questioning the underlying metaphysical assumptions on which physical science is based and exploring the fundamental meaning of basic physical-scientific concepts such as 'matter' and 'energy'. As well as providing an essential philosophical foundation for the developing field of 'New Science' and 'New Energy' research it also offers a fundamentally new paradigm of Fundamental Research. This is not experimental research in the ordinary sense so much as direct experiential research. Instead of being aimed at the gathering of quantitative data it is a form of qualitative 'phenomenological' research and investigation requiring no other instruments than the researcher's own organism. It is field research in the most direct sense – based on the researcher's ability to directly enter and experience the different fields of awareness and fields of reality of which our own organism is a living expression.

The framework of Fundamental Science and original methods of meditational or 'field-phenomenological' research developed by me over the last 30 years, are not based on the meditational practices of any established spiritual tradition, Eastern or Western. Unlike the majority of these practices, Fundamental Research makes use of the dyadic field – the combined awareness of two people – to amplify their mutual resonance with the inner field-states and field-patterns of awareness. It is these 'intensional' fields that constitute what I call the Inner Universe. Fundamental Research has not only its own unique meditational practices and methodological procedures, but its own methods of validation, based on comparing the experiences gathered by pairs of researchers.

Fundamental Science has profound theoretical implications for our understanding of the human as well as the natural sciences, fulfilling Marx's vision of a "human science of nature" which complements the "natural science of man". It also has direct practical applications in the fields of medicine and psychotherapy, offering as it does a fundamentally new understanding of the nature and meaning of health and illness. Fundamental Research has led to the development of new methods of therapy, transcending the division between somatic medicine and psychotherapy, and based on a direct organismic field-resonance between practitioners and patient.

Fundamental Misconceptions

The belief that subjectivity or awareness has its source in a localized subject or in particular localized objects that we are aware of (such as the brain) is one of a number of fundamental misconceptions challenged by Fundamental Science and Fundamental Research. Others include:

1. The misconception that action has its source in localized agents or selves, rather than being essentially the self-actualization of fields of potentiality.

2. The misconception that spin, velocity etc. have their source in particles, — when as Schrödinger put it, there is no such thing as a particle at rest, and that "properly speaking, one never observes the same particle a second time."

3. The misconception that 'power' is derived from sources of energy rather than energy having its fundamental source in power, understood as potentiality.

4. The misconception that gravity and light have their source in material bodies.

5. The misconception that capacities such as sight have their source in organs – as if the capacity to write could belong to an organ or instrument of writing.

6. The misconception that meaning, sense or significance is a function of language or other sign systems – when the latter are in fact all expressions of directly felt meaning or 'felt sense'.

7. The misconception that language has its source in patterns of experienced reality, rather than in the patterns of significance manifest in that reality.

8. The misconception that probability is a function of statistical variations from a random pattern.

9. The misconception that a given phenomenon can have its source in other phenomena, when in fact all experienced or observed phenomena are the manifestation of patterned fields of emergence.

10. The general misconception that the inwardness or intensional reality of phenomena – the inner universe – has its source in spatio-temporal relationships, when in fact all such relationships, including that of 'outside' and 'inside' are themselves but different types of extensional relation.

The Language of Fundamental Science

The Fundamental Misconceptions can be seen as expressions of a thinking enframed by a fundamental linguistic pattern dominated by the noun or noun phrase. Within the structure of every sentence, both Subject and Object are simple nouns such as 'energy', multi-word noun phrases such as 'electromagnetic energy', or complex noun phrases which include a verb such as 'the energy needed to drive a motor'. Scientific thinking rests, like scientific language, on the classification of nouns through modifying nouns or adjectives. Thus we speak of 'light energy' and 'heat energy', of 'potential' or 'kinetic' energy. Instead of asking what 'energy' as such essentially is, we consider only the relationship between different types of energy. We call this type of thinking adjectival thinking, operating as it does by classifying different types or properties of the same pre-given 'thing'. Adjectival thinking in turn is but one form of nominal thinking, in which the verb is subordinated to nouns and noun phrases. All action is seen as the action of a pre-given thing represented by the subject of a sentence, and serves only to represent its effects on another thing, – the sentence object – or to describe its properties. Nominal thinking is also nominalistic thinking, positing as it does a world of pre-given things represented as nouns, and making all action and interaction, all aspects and adjectives into functions or 'predicates' of those things. Even esoteric and mystical thinking fall in line with the same linguistic patterns – positing 'astral' or 'etheric' bodies in addition to the physical body, or other 'subtle' or 'etheric' energies in addition to physical ones. There is nothing wrong with classification and adjectival thinking as such, allowing us as it does to define and refine important distinctions in language. The basic patterns of language however, make it more difficult to think in a dynamic way in which the verb is not subordinated to the noun and its adjectives, action not seen as a function of things and their properties. The fundamental formula of the sentence is NP + VP. The equation reads: S > NP (noun phrase) + VP (verb phrase). This can be read as a proposition: 'things act' or 'nouns verb'. The assumption that accepted scientific propositions are representation of empirical fact ignores the entirely invalidated root propositions of language itself. Fundamental Science requires us to suspend belief in these root propositions which we so far have taken as fundamental, but which, far from being self-evident, have become a hindrance to understanding the fundamental nature of reality. Fundamental Science and Field-Dynamic Phenomenology go hand in hand with a dynamic thinking that gives ontological priority to the verb.

So far, the realities that could be described through this dynamic verbal thinking have only been suggested through mathematical equations such as $E = mc^2$. The equivalence of of matter and energy represented in this equation could however also be expressed in such terms as 'energy matters' or 'matter energies' – in both of which a word normally used as a noun is understood as a verb. And in this way, which take us back to the root meaning of 'energy' as formative activity (*energein*) it is no longer implicitly posited as some 'thing' which then acts in a formative way or is produced by some transformation. Instead we understand it as the very activity of forming and transforming, their essential dynamic. The turn away from nouns to verbal nouns (such as those which created through the gerundial ' –ing' ending in English) in turn has implications for our understanding of other basic scientific concepts such as 'attraction' for example. In conventional scientific language this concept takes the primary form of a noun and is thought nominalistically — as a 'thing' in the form of a 'force'. The nature of this thing is then differentiated adjectivally, so we can speak of different types of attraction — gravitational, electrical and magnetic. Here again however, the use of nouns such as 'attraction' implies some 'thing' corresponding to that noun, whereas the use of dynamic '-ing' words such as 'attracting', 'polarizing', 'charging', 'cooling' does not — and can all be thought of as different expressions of 'energy' as a formative and transformative activity, one which is not itself a thing or the action of a thing.

The only thinker ever to attempt to deploy a new dynamic language based on the gerund or verbal noun was Martin Heidegger. His language is so hard for many people to fathom, but in order to fathom the nature of fundamental reality, Heidegger not only ceased to privilege the noun but actively fashioned a new language which gave precedence to the verb and in this way expressed a new and dynamic way of thinking. Scientific propositions are not representations so much as translations of empirical research, framed by nominal language and nominalistic thinking. Retranslated in a dynamic way, a scientific proposition such as 'bodies exert a gravitational force of attraction on one another in space that causes them to move towards one another' could read 'gravitating is a gathering of space that weights on the bodying of movement'. Many such different dynamic retranslations are possible. Which of them would be most adequate as a formulation of fundamental truth would be the question addressed by Fundamental Research.

Though the different formulations might appear as gibberish to the layman, so also do the complex mathematical formulations of fundamental truth offered up by physics. And just as mathematical equations of current science must be proven and experimentally validated, so must the linguistic propositions of a future, more Fundamental Science, be selected according to their adequacy in giving expression to fundamental field-dynamic relationships. This is impossible as long as one pre-given pattern of relationships is taken as self-evident, simply because it is the basic pattern of everyday language. As Heidegger pointed out "Physics as physics can make no assertions about physics. All the assertions of physics speak in the manner of physics. Physics itself is not a possible object of physical experiment." The same is true of all the sciences, in so far as their assertions are framed within a specific language or universe of discourse which is not itself questioned.

The Nature of Physical Science

What we understand today as physical science is fundamentally flawed — is faulty. It is faulty because it defaults on five critical counts in its attempts to provide an account of the fundamental nature of reality

1. The first default is its inability, in principle, to explain how awareness emerges from an otherwise non-aware universe of matter and energy.

2. The second default is its inability, in principle, to explain how meaning emerges from an otherwise meaningless universe of matter and energy — except through patterns of significance imposed by a divine or human subject separate and apart from that universe.

3. The third default is its inability, in principle, to predict the form of even the simplest material units such as atoms and molecules from the probabilistic mathematics of 'quantized' physical energy fields.

4. The fourth default is its inability, in principle, to provide an account of what 'matter' or 'energy' fundamentally are. The most fundamental of scientific terms such as 'force' and 'energy', 'magnetism' and 'electricity', 'mass' and 'acceleration' are defined circularly — in terms of one another, and represented as a set of mathematical relationships to one another.

5. The fifth default: the inability of the physical-scientific world view to account for the qualitative dimension of even the simplest of experienced phenomena such as colour, or to account for the nature of sensory perception in all its forms.

Despite its claims to a solid 'empirical' basis, the physical-scientific world outlook denies the authority of the senses. For on the one hand it regards sensuously perceived phenomena as products of the brain — and in this sense no more than illusory 'effigies' of fundamental reality. On the other hand it explains the very process of perception as one in which physical sense data are passively received from these effigies, now taken as physically 'real' in a fundamental sense.

The Nature of Phenomenological Science

Physical science assumes a world of pre-given entities such as 'energy', 'matter', 'electromagnetism', 'light' etc. independent of our own awareness of them, and governed by a pre-given order that is also independent of the ordering or patterning of consciousness itself.

Phenomenology has traditionally challenged this assumption, recognizing that the known universe is the universe as we are aware of it, the universe of our current – and possibly quite limited – human awareness, as this is currently patterned or configured.

From a phenomenological perspective, since awareness of subjectivity is the very condition for our awareness of any phenomena whatsoever, it cannot itself be reduced to one phenomenon among others – or to a mysterious 'epiphenomenon' that is inexplicably and miraculously generated by an otherwise non-aware universe of matter and energy.

Phenomenology is the philosophical bridge between physics and psychology, recognizing as it does that our awareness of a world of external 'physical' phenomena in space-time is no less 'psychical' in character than our awareness of internal 'psychological' phenomena such as thoughts and feelings.

Twentieth-century phenomenology failed in its noble goal of establishing a new philosophical foundation for the sciences because it seemed to suggest that objects of consciousness were mere constructs of a conscious subject with no independent reality whatsoever.

Just as physical science still cannot explain what 'matter', 'energy' etc. fundamentally 'are', so phenomenological science had no answer to questions about the fundamental nature of 'things in themselves', the reality behind experienced phenomena.

Neither physical science nor phenomenological science, as currently understood, therefore constitute Fundamental Science.

Field-Dynamic Phenomenology

Fundamental Science marks a radical departure from twentieth-century phenomenology and physics.

Twentieth-century physics still tended to see phenomena such as charge, spin, momentum etc. as the properties of pre-existent particles such as electrons.

Similarly, twentieth-century phenomenology still tended to treat awareness as the property of a pre-given subject or ego (albeit one that was in some way 'transcendental' since it could itself never be turned into an object of consciousness.

Fundamental Science is 'field-phenomenological science', not based on a 'transcendental' phenomenology but on a field-dynamic phenomenology. Its foundation is the understanding that awareness or subjectivity itself is not the property of a pre-given human subject, or reducible to the features of human subjective awareness. Rather than being the property of any localizable subject of consciousness, human or otherwise, subjectivity or awareness itself has an essentially non-local or field character.

The terms 'physics' and 'physical' have their roots in the Greek verb *phuein* – meaning to 'arise' or 'emerge'. In its field character, awareness is not just the condition for our own conscious human experience of any physical phenomena or universe whatsoever. It is the basic field-condition for the very emergence of those phenomena in the first place.

From a field-phenomenological perspective, all observed or experienced phenomena are the manifestation of a primordial field or ground state of awareness, which can never be reduced to or explained by any of the phenomena that manifest or emerge from it.

To reduce any field of awareness to a product or function of any phenomena that manifest within it is like reducing dreaming – the open and ever-changing field of our dream awareness – to a product or function of some particular thing or things that we happen to dream of.

The First Fundamental Principle

The first basic 'phenomenological' principle of Fundamental Science is that no *field* of awareness can be reduced to or explained by the experienced phenomena or observed patterns of events that manifest or emerge within it.

The Second Fundamental Principle

The second basic principle of Fundamental Science is that no phenomenon or pattern of events can be reduced to or causally explained by other phenomena or patterns of events manifesting *within the same field.* To explain one phenomenon or pattern of events by another is equivalent to trying to explain how one dream object or event is 'caused' by another – when in fact both are expressions of the same source field of our dream awareness.

The First Fundamental Distinction

Unlike current phenomenological or physical science, Fundamental Science, based as it is on a field-dynamic phenomenology, recognizes a fundamental distinction that neither physics nor phenomenology have so far acknowledged.

The First Distinction which Fundamental Science makes is the distinction between, on the one hand, any specific pattern of events we are or could become aware of, and on the other hand, specific patterns of awareness as such.

All experienced or observed phenomena are patterns of events – patterns which we either recognize as stable objects of perception or represent as atomic and sub-atomic processes.

These physical patterns of events however, can only manifest as experienced or observed phenomena within fields of awareness that are themselves patterned in a way that shapes these experienced or observed phenomena.

In addition, however, it must be recognized that any experienced or observed pattern of events implies the possibility of other contrasting or complementary patterns. It is not just that the experience or observation of a given pattern is conditioned by specific field-patterns of awareness. Its very manifestation or actualisation is also the expression of a field of unmanifest or potential patterns.

Potential patterns of events are not reducible to actually observed or experienced patterns. Nor is the field of such potential patterns reducible to a spatio-temporal field – the field of emergence (phuein) or 'physical' actualisation.

Potential field-patterns, do not occupy physical space-time, nor are they reducible to 'energetic' fields plotted mathematically in any extensional dimensions of space-time. They have their reality only in fields of awareness.

The Second Fundamental Distinction

The second distinction which Fundamental Science makes is between extensional spatio-temporal fields and field-patterns on the one hand, and non-extensional or intensional fields on the other – the latter consisting of potential field-patterns and intensities of awareness.

What in physics now goes by the name of 'vacuum' or 'virtual' energy, 'ground state' or 'zero-point' energy, is neither a property of extensional space, nor a form of energy or matter 'filling space'. Its essence lies in no actual energetic patterns or material particles, nor even in 'potential energy' but in the formative potentials of intensional fields, and in intensities of formative potentiality.

What we think of as 'energetic' field-patterns and field-dynamics in extensional space-time are the dynamic self-actualisation of potential field-patterns of awareness which have a fundamentally non-extensional or intensional character.

The Fundamental Nature of 'Energy'

The search for new and more fundamental particles has now partially given way to the search for the source of a new more fundamental energy. This in turn must be founded on a new and more fundamental understanding of the nature of 'energy' as such – one which not only relates different types of energy or seeks to account for their localized effects, but addresses the fundamental question of what energy as such fundamentally is.

The term 'energy' derives from the Greek verb *energein* – which did not mean a potential or capacity for 'work' but rather a formative activity arising from formative capacity or pattern-giving potential – like the formative capacity of a potter to shape clay, the potter's awareness of its formative potentials and the formative activity or 'work' through which the potter actualises these potentials.

The idea of a closed universe containing a finite quantity of 'energy' is, from this point of view, fundamentally flawed. For the actualisation of any potential pattern or form does not reduce but increases the number of potential patterns or forms. Just as one idea 'leads to another' – or rather to multiple other ideas – so does all action multiply the possibilities of action, not only actualising a potential pattern or form but generating new potential patterns or forms. The intensional-field of potential patterns is not a static Platonic realm of Ideas or Forms but a dynamic field which is constantly expanding itself in and through its self-actualisation.

Action as such is not the 'work' of a pre-given force or agent, natural or divine. Instead it is the autonomous self-actualisation of a field of potential patterns which adds to itself through its actualisations. 'Energy' understood as formative activity (*energein*) is therefore (a) essentially autonomous – self-originating – and (b) unlimited and self-increasing.

As Plato recognized long ago, pattern or form as such, however, is essentially mass-less. Nor does it possess 'energy' in the current understanding of that term.

We can pick up a round plate and measure its radius but we cannot pick up or measure the roundness of that plate. Nor can its roundness as such possess any capacity for work. The roundness however, is the expression of 'energy' understood in the fundamental sense of *energein* – the formative activity that actualises potential field-patterns of awareness.

Just as form as such does not have its source in matter, so does formative activity as such not have its source in the 'work' of localized agents, such as a human potter. The human form and human capacities of the potter are as much an expression of energy as formative activity as the pot itself. Both potter and pot are a medium for the self-actualisation of a field of potential patterns.

Unlike actualised physical forms such as pots and pans however, potential patterns, like potential pots and pans have reality only within fields of awareness – such as the human field of awareness of the creative potter. That is why all intensional fields are essentially fields of awareness as such, and all the potential patterns they contain are in essence patterns of awareness as such.

The actualisation of any given pattern from a source field of awareness has dramatic results. For any actualised pattern or shape, form or figuration that emerges from a field of awareness, will, simply by virtue of being a form or figuration of awareness, also in-form and configure its own shaped and patterned field of awareness.

Ideas or concepts, for example, not only emerge within our field of awareness. They also shape and form our awareness. They can do so only because they are themselves shapings or figurations of awareness. As such, however, every idea or concept is potentially capable of totally re-shaping and transforming our awareness – configuring an entirely new field of awareness that is shaped and patterned by it.

The Third Fundamental Distinction

The third distinction central to Fundamental Science is between source fields of emergence, and the potential field-patterns of awareness latent within them, and the subordinate experiential fields they give rise to. The actualisation of any given pattern or configuration of awareness from a source field of potential patterns immediately results in the configuration of a subordinate, experiential field of awareness shaped by that pattern of awareness. No source field or universe is ever manifest as a field-pattern within the subordinate experiential fields that emerge from it – let alone reducible to the phenomena experienced in these fields. As repositories of unmanifest patterns of events and experience, source fields themselves withdraw into manifestation within the subordinate fields that emerge from them – disappearing in the very process of manifestation or appearing only as voids or 'vacuum' states within these fields.

All experienced phenomena and patterns of events are the manifestation of fields of awareness already shaped by patterns of awareness. They give expression to these field-patterns of awareness, which in turn emerge from a pre- and trans-physical source field. Fundamental Science understands the entire physical universe itself as a patterned spatio-temporal field of awareness. Far from being a closed system, this outer universe is itself but one extensional and energetic expression of an inner universe — the potentialities latent within its own intensional source field of awareness. The atoms, molecules, cells and living organisms that make up the universe as we perceive it, are the phenomenal manifestation, within our own highly patterned experiential field of awareness, of patterns of atomic, molecular, cellular and organismic awareness.

The Fundamental Nature of 'Probability'

According to Rupert Sheldrake, all material forms or 'morphic units', from atoms to complex organisms are actively in-formed by what he terms 'morphic fields'. He also presents a highly significant account of the probabilistic nature of these fields, understanding them essentially as made up of multiple potential field-patterns. He goes on to explain how the form of any given morphic unit or material structure is automatically stabilized through self-resonance with the organizing field-pattern it gives expression to. The field of potential patterns is itself stabilized through self-resonance with the actualised patterns that emerge from it, making it more likely or probable that this particular pattern will be reproduced or actualised in the future – not only in a given morphic unit but in all other similar units. In his account of morphic fields as 'probabilistic' fields, Sheldrake is in fact presenting us with a fundamentally new understanding of the nature of probability as such. Probability is no longer defined in a purely mathematical way, as a quantitative statistical variation from a random state lacking any manifest order. Instead it is understood as a state of resonance between a morphic field and its manifestation in a morphic unit. For Sheldrake himself, the principle of morphic resonance provided a new explanation for (a) the known structural form of atoms and molecules – something which cannot be accounted for or predicted from their quantum structure alone and (b) for the development or 'morphogenesis' of complex organisms in particular, whose forms, whether simple or complex, can likewise not be predicted or derived from their genetic structure.

The theory of morphic resonance also explains the nature of what Sheldrake calls 'formative causation' – how it is that apparently localized changes in the structural form or behavioural patterns of morphic units (whether atoms or molecules, cells or organs, animals or human beings), can have a non-local 'knock on' effect on the structure or behaviour of similar units, without any energetic interaction occurring between them. From the point of view of Fundamental Science, however, Sheldrake's principle of 'morphic resonance' is one of a number of Fundamental Dynamics governing the relation between any potential field-pattern belonging to a source field of emergence, and its manifestation in a subordinate experiential field. These Fundamental Dynamics constitute the challenge of Fundamental Science to the laws of Thermodynamics regarded as fundamental within Physical Science.

The Fundamental Dynamics

- All phenomena are experienced or observed patterns of events, whether these take the form of patterned objects of perception, patterns of sub-atomic events or patterned movements and processes.

- Any actual pattern, however, implies other possible patterns (contrasting or complementary) and has its source in a field of potential patterns.

- Potential patterns are by nature unmanifest patterns – they form no part of any manifest universe.

- Any manifest universe is a field of actualisation of potential patterns. Potential patterns as such have no reality within these manifest universes or fields of actualisation. They have reality only within fields of awareness, as potential field-patterns of awareness.

- The actualisation of a potential field-pattern of awareness results in the creation of a patterned field of awareness.

- All manifest phenomena, as experienced or observed patterns, are the actualisation, within an already patterned field of awareness, of other potential patterns of awareness.

- All manifest universes or fields of actualisation are themselves made up of patterned fields of awareness, emerging from a source field consisting of potential field-patterns of awareness.

- Since any given universe or field of actualisation consists of patterned fields of awareness, no phenomenon manifesting within it is merely a 'result' or 'effect' of other phenomena manifesting in the same field.

- Nor can the universe or field of actualisation itself be reduced or explained by any of the actual phenomena, any experienced or observed patterns manifesting within it.

- Instead all the actual patterns experienced or observed within any manifest universe, emerging as they do within already patterned fields of awareness, are shaped by the specific field-patterns of awareness that make up that universe.

- Awareness forms itself into patterns. Fields of awareness are made up of potential patterns of awareness.

- Each of these potential field-patterns of awareness, when actualised, constitutes an independent consciousness – one whose specific field-pattern of awareness in turn configures its own patterned field of awareness.

- The patterned fields of awareness configured by any consciousness constitute an experienced or observed universe, its reality field.

- Reality frameworks are reality fields composed of multiple consciousnesses but sharing a common field-pattern.

- 'Energy' is the formative activity (*energein*) through which potential field-patterns of awareness become manifest as actual phenomena, within the patterned fields of awareness (reality fields) of different consciousnesses.

- As such it is also the autonomous self-actualisation of source fields of awareness within the patterned fields of awareness that emerge from them.

- But the actualisation of any potential field-pattern of awareness within a patterned field of awareness automatically multiplies the number of new potential patterns that can manifest as phenomena in that reality field.

- In this sense no universe or reality field is a closed system with a finite quantity of energy. For energy, as formative activity, not only has an inexhaustible source in potential field-patterns of awareness, but also multiplies itself in the very process of manifesting or actualising these patterns – generating new potential patterns.

- All action automatically multiplies the possibilities of action.

- Action is not the activity of a pre-given and localised agent – human, natural or divine, but the self-actualisation of non-localised source fields of awareness through which they actualise potential field-patterns of awareness.

- This formative activity is essentially autonomous and inexhaustible, multiplying its own formative potentials through their actualisation.

- Energy, as the manifestation of formative activity in already formed or patterned fields of awareness, is itself autonomous and inexhaustible – adding to its own potential.

- Potential energy and potential forms or patterns are the primary link between 'energetic' fields and fields of awareness, between fields of actualisation and the source fields of potentiality from which they emerge.

- In particular they are the link between the physical universe or reality field and other reality fields in which different potential field-patterns of awareness are actualised.

- Potential field-patterns occupy no physical space. They do not form part of any extensional space-time universe, but belong to non-extensional or intensional fields of awareness.

- Awareness is the inwardness of energy – its intensional source.

- Matter is the outwardness of energy – its extensional manifestation.

- Awareness is essentially awareness of potentiality.

- Consciousness is essentially awareness of actuality.

- Material form (*morphe*) is the expression of energetic field-patterns.

- The formed materialization of an energetic field-pattern automatically establishes non-energetic resonance with an underlying field-pattern of awareness.

- This 'morphic resonance' (Sheldrake) makes it more likely or probable that the pattern will continue to be actualised, thereby stabilizing the structure of its material forms.

- Probability as such is a dynamic relation between potential and actual patterns of energy and awareness, whereby the actualisation of any potential field-pattern makes it more likely or probable that the pattern will be maintained, whilst at the same time (a) generating alternate possible patterns which manifest in other reality fields, (b) attracting specific higher-order field-patterns into manifestation in the same reality field.

- Morphic resonance is a non-energetic resonance between (a) a potential field-pattern and an actualised one, and (b) a non-energetic resonance between actualised lower order patterns and potential higher-order ones.

- Morphic resonance is a form of non-energetic or 'formative' causation because it is essentially resonance between actual energetic field-patterns and potential or intensional field-patterns of awareness.

- Intensional fields of awareness are not made up of any experienced or observed patterns of events that we are or could be aware of within our own extensional field of awareness.

- Instead they are composed of patterned tones and intensities of awareness as such — familiar to us as moods of patterned tones of feeling of the sort expressed in vocal and musical tones.

- Tone and pattern are intrinsically related. Field-patterns of awareness are not simply arrangements of different tones and intensities of awareness. They possess their own fundamental tone. In the physical reality field this expresses itself as the resonant frequency of patterned material forms.

- Conversely, every tone bears patterns within itself — a fact that in the physical reality field is evidenced by the ability of particular sound frequencies to create distinct field-patterns in a fluid medium.

- Morphic resonance is also the expression of an intrinsic relation of resonance between form or pattern on the one hand (*morphe*) and tone on the other, a relation that takes the form of sound as tonal form or pattern.

- Tone re-sounds in formal patterns, just as formal patterns possess their own inner sound or resonance.

- Sounds as such are patterns or shaping of tones, just as tones are resonances of patterned shapes. Forms sound and sounds form.

- Communication of all forms is based on morphic resonance between a potential field-pattern of significance and its manifest expression — for example in speech or music. This resonance amplifies and 'broadcasts' the basic tonality or resonant frequency of that field-pattern in a way that can be picked up by others who are tuned to it.

- Physical energy is outward or extensional energy — the patterning of events within extensional, spatio-temporal fields, for example through oscillatory patterns or 'vibrations', each with their own frequency and amplitude.

- 'Etheric' or 'subtle' energy is inner or intensional energy — inergy.

- It is the self-patterning of different tones and intensities of awareness as such — giving rise to different field-textures and field-densities of awareness.

- Physical energy possesses characteristics of polarity and charge.

- Etheric energy possesses characteristics of tonality and intensity.

- Physical particles are the manifestation of discrete quanta of energy.

- Etheric particles are the manifestation of distinct qualia — qualitative tonalities and intensities of awareness.

- Field-patterns of awareness manifest firstly as etheric or inergetic patterns — that is to say as patterned tones and intensities, textures and densities of awareness.

- Energetic patterns are the outward manifestation of etheric or inergetic patterns.

The Fourth Fundamental Distinction

Outwardly sensed qualities of energy in the form of light, warmth, gravity, electrical charge, magnetic field-strength, sound vibrations etc. are at the same time quantitatively measurable. But each have their counterpart in felt inergetic qualities, none of which are quantitatively measurable. The radiance or light of another human being's gaze, their warmth or coolness, distance or closeness as human beings, their lightness or heaviness of mood, levity or gravitas, magnetism or lack of it, even their excitement or charge or flatness and lack of charge — all these are not measurable energetic quantities, nor even outwardly sensed energetic qualities, but inwardly sensed forms of inergy. The moment we study someone's eyes as an object we cease to perceive the luminosity of their gaze, an inergetic quality which may bear no relation to the measurable quantity of light reflected by their eyes. Measuring the temperature of a person's body is in no way the same as feeling their inner warmth as human beings. Nor are the feeling tones echoed in their tone of voice reducible to measurable vibrational pitches.

'Inner', 'intensional' or 'inergetic', tone and intensity, warmth and light, electricity and magnetism, density and gravity are in no way less real to our inner senses than their measurable energetic counterparts. Instead they are more immediately sensible and real to us than the latter, which often pass us by – or which our outer, extensional senses are unable to register. What Gendlin calls 'felt sense' consists of felt tones and intensities, densities and textures of awareness. It is thereby an attunement to those intensional field-patterns of awareness, which manifest both as energetic patterns and as material forms, as observed or experienced patterns of events.

Both mathematical and linguistic structures can be used to represent these intensional field-patterns. Field-patterns of awareness, however, find expression not only in language or mathematics but in patterns of experienced or observed reality itself. Both words and the phenomena they describe, both language and experienced reality, both mathematical equations and the pattern of events they represent, are expressions of these field-patterns. Sign systems such as mathematics and language do not merely represent or signify observed or experienced events. They are themselves a direct manifestation of the underlying field-patterns actualised in those events, and signified by them.

Language and mathematics can provide an account of the fundamental nature of reality if they are not only seen as representations of experienced reality or observed events, but also understood as expressions of felt sense — in resonance with underlying field-patterns which manifest in those events. A belief is an expression of felt sense — of inner or intensional reality — but one that is represented as a factual assertion about outer or extensional reality.

As long as we judge scientific or religious beliefs by their correctness as verifiable representations of outer reality, we ignore the fundamental role of scientific and religious sign systems in giving direct expression to felt sense and inner reality. As long as we confuse the accurate representation of outer reality and the outer universe through experience and observation with the accurate expression of felt sense and inner reality, inner reality will continue to be represented in terms distorted by outer reality - and vice versa. Only through the active and methodical deepening of felt sense can it be transformed into a felt resonance with the inergetic field-patterns of awareness that constitute the inner universe. Only then can language and other sign systems give direct resonant expression to the intensional field-patterns of awareness, actual and potential of this inner universe. For these field-patterns of awareness, and the qualitative tones and intensities, densities and textures of awareness through which we sense them are themselves felt patterns of sense or significance — patterns of meaning whose physical metaphor is matter itself.

The Fifth Fundamental Distinction

The fifth distinction central to Fundamental Science is between inner field-patterns of awareness, experienced through felt sense, on the one hand and their manifestation in experienced phenomena — in both, words and things, language and experience. Signs are what signify or point to senses. But meaning or sense as such is not a property of signs or sign structures but consists of felt qualities and field-patterns of awareness that are intrinsically meaningful. It is these that constitute what Gendlin calls 'felt sense'.

Semiotics as Semiophenomenology

Experienced phenomena, whether in the form of words or things, function as signs. Words have the character of formal signifiers, appearing to signify perceived things or to represent concepts of those things. Things themselves however, have the character of physical signifiers. They are material metaphors of those intrinsically meaningful qualities and field-patterns of awareness that constitute fundamental sense. Just as words are things – sounds and sound patterns – so are things themselves 'words' – expressions of sense. Both emerge as phenomena in our field of awareness. But field-phenomenological science is also a field-phenomenological semiotics that recognizes the sign character of all phenomena, their function as signifiers of felt qualities of awareness or 'qualia'. Meaning or sense is not something we attach to the universe through language and thought, nor is it only a property of formal signifiers such as words. Meaning quite literally matters – materializing itself in the outer universe. Conversely, matter itself is intrinsically meaningful – a materialization of meaning as felt sense. The fundamental distinctions between localized phenomena and the fields of awareness they give expression to, between energy or outwardly perceived qualities of things, on the one hand and 'inergy' or sensed field-qualities of awareness on the other, all find expression in the Fifth Fundamental Distinction – the distinction between signs and their felt sense, and the understanding of all experienced phenomena as phenomenal signifiers of felt sense. Felt sense or '6th' sense is an expression of a fundamental dimension of reality – the 5th Dimension. The 5th dimension is not a fifth dimension of outer space but the depth dimension of inner meaning or sense present, not just within words, but within all phenomena in their function as signs.

Future scientists, as Fundamental Scientists, will read our current scientific accounts of the atom and other material structures in a quite different way. They will read them not just as fallible representations of material structures and energetic patterns but as metaphors of our current self-structure and our own dominant field-patterns of awareness. The Fundamental Scientist today is still in the position of someone trying to explain to those who cannot read that the patterned ink marks they see on a page are the visible material manifestation of an invisible, unmanifest and multidimensional world of meaning. This is a world that cannot be revealed by the most exhaustive chemical analyses, X-ray spectroscopy or nuclear magnetic resonance scanning. To research it requires resonance of quite a different sort.

Fundamental Research, based on specific methods for establishing direct organismic resonance with inergetic field-patterns, field-states and field-dynamics will not replace but feed into conventional physical-scientific research. For its results will offer new insights into energetic field-patterns, field-states and field-dynamics. The exploration of different worlds or reality fields through inner or intensional space will complement their exploration through outer space — and offer new gateways to it.

There is no more dangerous myth than the idea that meaning and truth is a property of the word — of religious or scientific assertions about the world — rather than something intrinsic to our immediate wordless awareness of the world. It is the felt directions, tones and textures of this awareness that constitute the essential sense of both words and of the worldly phenomena they describe or refer to. Sense is not a property of the words we use to refer to and describe phenomena, nor is the truth of words a function of their correctness in representing these phenomena in thought.

The sense of a dream, poem, painting or piece of music is not its reference to some outward physical phenomenon or even its representation of some inner subjective feeling. It is rather the immanent sense or meaning of the fact or feeling itself that it seeks to bear across. Its truth does not lie in the accuracy of its representation of that fact or feeling but in its resonance with the intrinsic sense borne across by it.

The meaning or 'sense' of a poem does not lie in the fact that it refers to some phenomenon, nor is its truth a function of the way it represents that phenomenon – a landscape, say. On the contrary, the meaning it conveys is intrinsic to the phenomenon itself. It is the meaning immanent in the landscape as such that is borne across wordlessly to the poet's awareness and borne across too, by the wordless senses of the poet's words. But even the most prosaic description of an everyday event or experience in words has a sense that we understand wordlessly – because it derives not from words but from the event or experience itself – from the world.

Meaning is immanent in our immediate lived awareness of the world, both as a world of things with a pre-assigned practical significance within an overall world-structure or 'scheme of things', and as *phenomena* with a deeper significance – one that 'shines forth' (*phainesthai*) through them. The sensed significance of phenomena is not their pre-assigned place in a spatio-temporal world structure or scheme – an already patterned field of awareness. The sense or meaning of a sign is not some 'thing' it 'refers', points or direct us to in this world scheme (as a road sign points us to a destination, or a weather sign points towards rain). This is merely its signification within a pre-assigned scheme of things. Its sensed significance is not something it directs us to but a direction of awareness to which it draws us. The root meaning of the word 'sense' is a way or direction. Sensing the significance of something on a deeper level means letting our awareness be directed by it in a new way.

A physical symptom, for example, may be taken as a sign in the first sense — referring or pointing to some 'thing' such as a possible organic disease. Alternatively, both the sign and the supposed 'thing' to which it points may be taken as something that points us in a certain direction — directing our own awareness towards a felt inner sense of dis-ease. Similarly, a dream symbol may be taken as a sign of some other 'thing' — a repressed drive or desire for example. But in interpreting it in this way, the psychoanalyst, like the physician, is merely locating a sign within a pre-conceived structure of signification — treating one sign as a pointer to another. When the analyst makes an interpretation of a dream symbol, or when the analysand 'free associates' around their own dreams, what they come up with are but further symbols in a chain of signification. When the physician diagnoses a patient's symptoms as a sign of some medically labelled disease, all that is happening is that one sign has been 'explained' as the effect of another. The question of what the organic disorder itself may be a sign of is reduced to a question of its physical 'causes'. Reducing sense to signification is like 'explaining' the existence of a specific word in a sentence as an 'effect' of previous words, or reducing the meaning of this word to its relation to other words in the overall structure of the sentence.

The sensed meaning of a poem, painting or piece of music can in no way be reduced to the words with which we represent it or to some 'things' it depicts. Similarly, the sensed significance of a dream symbol or body symptom, indeed of any phenomena we may be aware of, by contrast, is not reducible to some other sign or set of signs, whether these take the form of words used to represent its significance or some pre-conceived 'thing' that these words are taken to signify. It is something that is felt directly in our awareness as a qualitative dimension of awareness to which it draws or directs us. The sensed significance of patterned musical tones for example, lies in the patterned tonalities of awareness they evoke. Both, the sensed significance of words and the inner meaning of musical tones lies in their wordless inner resonance. Indeed the same can be said regarding the inner meaning or sensed significance of all phenomena we experience in our sentient field of awareness. The infant does not hear a sound as that of a 'car passing by' or 'a clock ticking', for cars and clocks are nameable 'things' only by virtue of their place within a pre-assigned structure of signification — the consensual reality in which we dwell as adults. For the infant on the other hand, what the thing itself essentially is, is the way it resounds within their awareness.

Sensed significance or sentience, inner or felt sense, is not intellectual or practical signification but inner sound or resonance — felt resonance. Resonance, not reference or relation, is the essential relation of sound and sense, 'signifier' and 'signified'. This is not merely a phenomenological or psychologistic hypothesis but a physical–scientific one and a fundamental scientific truth. For as cymatic research has visibly demonstrated, material structures are themselves stabilized patterns of vibration — of inner sound, patterns which can be altered and transformed by the use of specific sound frequencies. Even the individual vowel sounds of speech, spoken into a tonoscope of the type invented by Hans Jenny and allowed to vibrate a powdered surface, produce clear geometric patterns similar to ancient mandalas. Resonance itself is essentially morphic resonance — a resonance between actual and potential pattern or form (*morphe*). Actualised extensional patterns are stabilised by intrinsic 'morphic' resonance with the intensional or potential patterns that are their source.

The Sixth Fundamental Distinction

The space of our felt resonance with the spoken word or a piece of music is not the outward, extensional space in which sound vibrations are transmitted by molecules of air. It is an inner semiotic space, a space of felt meaning or 'sense'. The sixth distinction on which Fundamental Science is founded is between outwardness and inwardness as such. Specifically it distinguishes between an extensional understanding of inwardness − the interiority of an extended three dimensional space or container such as a room or box, and inwardness conceived in a different and non-extensional way. The clue to the nature of non-extensional inwardness, and thus to the very nature of intensional reality and intensional fields, lies in the inwardness of the word or of any signifier. The inwardness or inner dimensions of the word − its meaning − is nothing extensional in character. Nor is it a finite interiority, bounded or limited extensionally. This distinction between an extensional inwardness and an intensional one — between outward inwardness and inwardness as such − is fundamental to the very concept of an inner universe distinct from an outer one.

The Nature of Fields of Awareness

Where do you see a clock? On a mantelpiece, for example. That is where you see the clock. But where do you *see* the clock? What is the space of your *seeing* as such? We see clocks on mantelpieces, the stars in the heavens. But the space of our seeing as such is nothing localizable like a flower in a field. It has a non-local or field character — being the field of visual awareness. We see trees in a wood, and each of them has its place in a spatial field. And yet there is a sense in which we never see the wood for the trees, for in focusing on any particular tree or group of trees we cease to see the wood. Even if we take an aerial photograph of the wood, we don't see the wood for the wood — for now the wood as a whole is like a single tree in a larger wood, a larger visual field of awareness that we do not 'see' as such.

The physical–physiological model of vision would like to have us believe that 'seeing' is a result of light waves striking the retina and producing signals transmitted to the brain via the optic nerve. But if seeing is something that takes place in our heads, and visual images are something produced by the brain and 'projected' outwards as apparent objects in space then 'where' exactly is 'space'? Is it in our heads? Surely not, for our heads are also objects in space. And what or where are the objects from which light waves are supposedly radiated or reflected in the first place - if all we know about them comes from perceptions or sensations generated by the brain? Last but not least, there is an inherent circularity of paradox at the basis of brain science: namely that from its own perspective, the brain itself - perceived as a spatial object of perception – is nothing more a *figment* of the brain. Upon this paradox the whole explanation of visual perception on the basis of brain science breaks down. For the brain itself is nowhere to be found except as a figment of itself!

The many paradoxes and contradictions in the physical-physiological model of vision apply to other modes of sensory perception also. We hear a car passing by on the street. But where do we hear the car? What is the space of our hearing as such? When we are listening to a piece of music there is a sense in which we are separated from a sound source in space, and there is another sense in which there is no spatial separation at all — we are within the music and it is within us. Our appreciation of the music is not a result of sound waves striking the eardrum and producing nerve signals in our brain that then trigger physiological changes which we experience as subjective emotions. For were we not already attuned to the music — within its tones and chords — its physical expression as sound would have no effect on us at all, no matter how high its volume and how tangible the bodily vibrations it produced in us. The space of our inner resonance with a piece of music, just like the space of our understanding of a person's words, or the space of our seeing — is nothing spatial in the ordinary sense — it has no measurable extension.

At the centre of Fundamental Science is the concept of non-extensional spaces and non-local fields of awareness. The idea of a non-extensional or 'intensional' space seems like a contradiction in terms for is not space as such defined by extension? And yet what appears, at first sight, as an affront to common sense is the most fundamental dimension of our common human experience. Seeing is not something that occurs in space but the configuration of a spatial field of visual awareness.

The so-called 'subjective' dimension of our human experience is not limited to our own idiosyncratic mental and emotional responses to perceived outer reality. It constitutes a fundamental dimension of that outwardly perceived reality, one not less but more real than what we currently understand as the 'objective' dimension of experience. Fields of awareness are the very condition for the perception of any localized object by any localized centre or subject of awareness. But even though our field of visual awareness may stretch to the heavens, our field of auditory perception to events taking place miles away, these fields of awareness, though giving extensional spatial form to our perception, are themselves nothing extensional – nothing localizable in space. We would be quite wrong therefore to consider the physical–scientific notion of energetic fields as something essentially to do with extensional space. For energy fields reveal themselves only through their local effects in space. And what does Einstein's theory of General Relativity tell us if not that extensional space itself is nothing absolute? It is not a pre-given space within which bodies move but the field-pattern of their movement, one whose shape or configuration is altered by that movement at every point in time.

The concept of fields of awareness is a Fundamental Concept, a foundational concept of Fundamental Science. So too is the concept of non-extensional or intensional space. Extensional space opens up within fields of awareness that are not themselves localizable in extensional space. What appears as the extensional spatial field of our sensory awareness has itself fundamentally non-extensional reality and opens up within a non-extensional or intensional space. But what sort of character can such a non-extensional space possible have?

Once again, is this not an offence to both science and common sense? Here we come up against one of the most stubborn prejudices of human thought and language, a prejudice that itself constitutes the greatest offence to common human sensibility. This is the prejudice that when, for example, we talk of being 'close' to someone or sense that they are 'far away' we are merely employing spatial metaphors. The proposition at the heart of this prejudice is a simple one: that the distance we feel to another human being is less 'real' than the extensional, spatial distance separating us from them. Authors such as Lakoff and Johnson have pointed out how full language is of spatial metaphors — feeling 'close' or 'distant', 'uplifted' or 'weighed down', 'holding' something in one's mind, being 'in touch' with someone or something etc. Their conclusion restates the old prejudice that such metaphors are second hand verbal borrowings — derived from our primary 'objective' reality as physical bodies in extensional space and applied to a secondary, 'subjective' dimension of reality — our reality as feeling beings.

This linguistic misconception immediately forecloses the possibility of a genuinely scientific understanding of human subjectivity itself — not as a mere by-product of bodies in extensional space and time but as the human expression and experience of fundamental reality. It prevents us from grasping scientifically that fundamental reality is intensional reality consisting of intensional space, intensional time, intensional field-patterns and intensional energy. That the type of inner closeness or distance we feel to other human beings is therefore not less but more real and fundamental than extensional closeness and distance — our directly felt experience of intensional space. That the inner warmth we feel radiating from another human being is not less but more real and fundamental than the measurable temperature of their body. That the inner light radiated by their gaze is not less but more real and fundamental than the measurable light energy reflected off or focused by their eyes in extensional space. That talk of a person's felt inner qualities of 'light' or 'darkness', 'warmth' or 'coolness', 'lightness' or 'heaviness', 'levity' or 'gravity', 'magnetism' or 'electricity' is not indirect metaphorical thinking but a direct felt sense of fundamental reality – fundamental reality in the form of intensional energy or 'inergy'. The foundation stone of Fundamental Science – its 'philosopher's stone' – is the acknowledgement of intensional reality as fundamental reality, a reality composed not of localized material objects or localized subjects or egos but of fields and field-patterns of awareness of subjectivity.

Descartes posited only one type of reality besides extensional reality (*res extensa*) and that was the reality of the knowing subject, ego or 'I' (*res cogitans*). Hence his famous maxim *cogito ergo sum*. Extensional reality constitutes an object of cognition for the cognizant subject or 'I'. What he did not recognize was the field character of subjectivity itself, and with it, the nature of extensional bodies, not as objects for a subject but as phenomena manifesting within fields of subjectivity of awareness. Field-dynamic phenomenology puts us in a position to consider the nature of a body – any body – as a boundary state between outwardness and inwardness, extensional and intensional space. The nature of an extensional body in other words, is nothing essentially extensional. It consists of a surface envelope, one which is at the same time a dynamic interface between two fields – a bounded spacetime field of extensional reality and an unbounded inner field of intensional reality. Since all physical bodies emerge (*phuein*) from a primordial field of awareness, they are essentially figurations of awareness. The surface envelope or boundary is thus essentially an envelope or capsule of awareness – not human ego awareness as we know it, but a type of fundamental awareness – a form of natural pre-egoic awareness from which all other, more sophisticated forms of animal and human awareness arise. The 'size' of an awareness envelope, by its very nature is something impossible to measure in purely extensional terms. Our own outer field of awareness itself extends beyond our own body to include and embrace all other bodies in our spatial environment. From the point of view of others however, its boundary is marked by the boundaries of our body. Our physical body is our own envelope of awareness as this is perceived within the outer field of awareness of other beings or 'awareness units' within their own envelopes of awareness.

The Fundamental Void

Understanding the intensional character of fundamental reality fills a void left by physics and philosophy, physiology and psychology. For it is in place of the fundamental concept of intensional reality, intensional fields and intensional space that psycho-physical dualism arose, together with the dichotomies of subject and object, matter and consciousness, idealism and materialism, and more recently, positivism and phenomenology.

The identification of reality with extension is rooted in the earliest eras of Western thought. To begin with extension was equated with discrete extensional bodies or with a continuous extensional medium — whether fluid or airy or ethereal. It was the Greek 'atomists' who first replaced the ontological duality of Being and Non-being with a cosmological duality of extensional bodies on the one hand and empty space on the other. But Aristotle denied even the possibility of a spatial vacuum free of extensional bodies or an extensional medium, as did Descartes centuries later. And Newton himself could never fully accept the idea that gravity represented action at a distance, unmediated by some form of subtle medium, material or immaterial. Like the Stoic philosophers, Newton's first picture of the cosmos was a stellar realm surrounded by an extra-cosmic void — a realm of pure chaos. To accept the notion of a void or vacuum seemed, as it did to countless previous generations of thinkers, intrinsically unnatural - in a way summed up in the maxim that 'Nature abhors a vacuum'. It also seemed irreligious, for through the identification of created reality with extension, the concept of a vacuum implied also an absence of God within the cosmos. Religious accounts however, themselves posited a type of creation 'ex nihilo' - a nothingness from which the extensional world was created. Fundamental theological as well as philosophical and scientific issues all centred, therefore, on what sort of reality could be conceived beyond a created world or even an eternal and uncreated cosmos if the latter was essentially an extensional space-time continuum. The only answer that seemed conceivable was some sort of extra-cosmic void or intra-cosmic vacuum, a nothingness out of which God created the world or, as in Eastern philosophies, a void which itself gave birth to the world.

Today, science itself has no difficulty accepting the idea of a void — a spatial vacuum or energetic 'ground state' possessing extension but not filled with a continuous extensional medium or ether. For Einstein's General Relativity seemed to make the idea of such an ether unnecessary to account for gravity - replacing it with the notion of curvatures of space itself. Yet that still left the question of what lies beyond extensional reality, now conceived as a shape of space in time. On the other hand science remains completely closed to the logical inconsistency — so obvious to the philosopher — of postulating a specific time at which time itself 'began'. And yet if the entire space-time universe is at the same time thought of as having a specific spatial shape — a torus or doughnut shape for example — then what sort of 'space' or 'space-time' lies beyond its boundaries?

At the same time light and gravity have become more of a mystery than ever, the one defining the curvature of space, the other existing as waves filling the vacuum of space and not ultimately dependent on the existence of sources of gravity in the form of extensional bodies. The puzzle of 'dark matter' — the fact that 90% of the gravitational mass of the universe remains unaccounted for — indicates that physics still has a long way to go in attempting to fill the voids in current cosmologies, let alone explaining what they essentially are, and from whence they derive.

Historically, the universe has been conceived of as finite or infinite in extent, continuous or broken up into atoms and collections of atoms separated by empty space, surrounded by an extra-cosmic void or filled with intra-cosmic vacua, flat or curved, spherical or toroidal. Yet the recurring problem, and the one that cannot be gotten round, is that the origins of an extensional space-time universe (or of multiple space-time universes) cannot, in principle, be conceived or mathematically accounted for in terms of space-time, i.e. in purely extensional terms.

Yet what appear as abstruse or intractable problems of physics and philosophy, mathematics and theology are answered by everyday human experience. Aristotle defined a vacuum as a space in which the presence of a body is possible but not actual. And we all know of a non-extensional space which fits this definition − the space of our imagination. We all know too, of an extensional space with no actual extension in space − the spaces of our dreaming. To which we may add the space of our seeing and hearing, which forms no part of any space we see or any objects we hear in it. Perhaps it is not surprising therefore that it was not a physicist but a psychoanalyst − Donald Winnicott − who first ventured a description of non-intensional space, not as an empty vacuum but as 'potential space'. For in each moment of time and in every situation in life we are aware of a psychic field of different possible or 'potential' actions or decision we can take, words we could utter, or movements we could make in space − none of which themselves have any extension, take up any space or are visible in (extensional) space.

It was the Greek thinker Heraclitus who first pointed to the non-extensional nature of the psyche itself, and characterized the essential nature of intensional space as unbounded interiority. The term 'psychology' derives from the Greek words *psyche* and *logos*, and Heraclitus's saying – being the first to conjoin these two words in a single saying, can truly be considered as the founding statement of any fundamental 'psycho-logy'.

"You shall not know the limits of the *psyche*, no matter how far you go about it, so deep is its *logos*". The term 'in-tension' means to tend inwards. The inwardness of the *psyche* is an inwardness of its word or *logos*. This is not an ordinary spatial insideness but an unbounded interior space of meaning or inner resonance. 'Meaning space' is that intensional space that constitutes the true inwardness of any text. It must be emphasized again, however, that this meaning 'space' is not 'space' in any merely metaphorical sense. On the contrary, it is a more fundamental or primordial space – understood as a space of awareness latent with potential meaning or sense.

This, however is where attempts to persuade scientists of the reality of intensional fields of spaces of awareness and meaning founders. For it is like attempting to persuade people who have *never learned to read* that within the apparent three-dimensional boundaries of a book there lie invisible but unbounded dimensions of meaning of a sort that no amount of 'scientific' analysis of the printed ink marks on a page or pixels on a screen of text will ever be capable of providing 'evidence' of.

Fundamental Linguistics

In so-called 'pre-scientific' eras of Western as well as Eastern thought, the process of 'emergence' (*phusis*) of natural phenomena was compared with the process by which the words themselves emerge and take shape in our awareness. Nature as such was understood as a living language, a book of life — an inscription or encryption of divine 'speech' (*logos*). And indeed, no better example of the Fundamental Dynamics can be found than in the process of speech formation. In the process of speech formation, one word does not 'grow' into another, or 'develop' into a multi-word phrase or sentence. The words we choose to express ourselves are selected from a field of possible words, and each choice implies the possibility of other choices, other possible words. The selection of any given word from this field of possible words, however, automatically limits the field of following words from which we can draw. Thus, following the word 'think', we would expect words such as 'of' or 'about' or 'carefully', just as following the word 'scientific' we would expect words such as 'research', 'theory' etc. Successive phenomena, as events of emergence, are no more linked by causal chains than succession of words in a sentence. The use of the verb 'think' does not 'cause' us to follow it with the word 'of', for example. What happens instead is that a given word attracts following words from a higher-order field of multi-word phrases such as 'think of' or 'think about'. It is an example of what Sheldrake calls a 'morphic germ'.

In general, any linguistic unit, whether a single sound, word, or multi-word phrase, can act as a morphic germ or 'attractor' for other words which go together or 'collocate' with it as within a field of higher-order units, including not only multi-word lexical and syntactic phrases but whole clauses and sentences.

The role of linguistic fields and field-patterns in the genesis of linguistic forms therefore exactly parallels Sheldrake's understanding of the role of 'morphic fields' and field-patterns in the process of 'morphogenesis' — the genesis of biological forms — and with the same consequences. For according to Sheldrake, the spontaneous emergence of new biological structures or behaviour in a given organism automatically makes it more likely or probable that these will be reproduced, not only in the organism in which they appear, but in other organisms of the same type — independently of any biological or energetic interaction between them. Once again, language provides us with the best and most evident example of this process of 'formative causation'. The fact that a new word or phrase is used by someone to express a given meaning does not 'cause' them to use it again or 'cause' others to use it, but it does make it more likely or probable that they will do so. This cannot be explained simply by the fact that others read or hear that word or phrase used — for whilst this may be a necessary condition for their use of the word it is not a sufficient condition. The new word or phrase is adopted through morphic resonance — because its form (*morphe*) carries a particular resonance that is in tune with felt senses or meanings that would otherwise lack verbal expression, not only for the initial user of a term but for others too. It emerges from a field of potential words and phrases which in turn give expression to hitherto unmanifest field-patterns of significance.

The structural patterns of language — phonological, semantic, lexical and syntactic — are the object of detailed analysis in the science of linguistics and represented through the most complex of formal schemas and 'meta-languages'. But the linguistic field-patterns from which we draw in the process of speech formation cannot, by their very nature, be distilled from actual examples of language use, for they include not only actual but potential usages, words and word-patterns. If they did not, language could not evolve. Nor can these field-patterns be represented in a scientific meta-language — for that language is itself but one expression of a field of potential patterns, and will necessarily impose its own actual pattern on the 'object language' studied. The field-patterns of language, which include everything from the sound patterns of words to complex, textual patterns, are by nature irreducible to their actual expression in ordinary language use or linguistic terminologies and meta-languages. They are the very field-patterns of sense or significance that both language and experienced reality itself give expression to. Once expressed in any sort of linguistic form however, certain of these patterns are stabilized by self-resonance and linguistic structures, like material or biological structures, becoming something whose structures can be represented scientifically. At the same time, the very emergence of particular linguistic patterns allows new potential patterns of significance to be expressed linguistically. Hence the basic paradox of any given language - which on the one hand appears to be made up of a finite number of sounds, words and structures — and on the other hand offers seemingly unlimited potentialities for creative expression.

This is so because language as such emerges from a source field, which includes countless potential field-patterns of sense and significiation as well as actual ones. And the emergence of the actual from an intensional source field of potentials automatically adds to and multiplies those potentials. "All action increases the potentialities of action". (Seth) This is yet another illustration of the Fundamental Dynamics. Conversely, however, the Fundamental Dynamics can been seen as an expression of a Fundamental Linguistics of the Inner Universe, understood *phenomeno-logically*, as the speech or *logos* that finds expression in *phenomena*.

The Seventh Fundamental Distinction

The seventh distinction acknowledged in Fundamental Science is the distinction between physical phenomena on the one hand and primordial phenomena on the other. The word 'phenomenon' is rooted in the Greek verb *phainesthai* – to 'shine forth' or 'come to light'. A primordial phenomenon is a phenomenon in the primordial or fundamental sense – that which shows itself, shines forth or comes to light through its physical manifestation or emergence (*phusis*). Light itself, understood as a primordial phenomenon, is not simply electromagnetic energy in the form of physical light. It is the light of awareness – without which no physical phenomena, including physical light and its properties, would be visible, and without which no phenomena could appear to us in any form whatsoever. As physical phenomenon, language consists of ink marks on a page, pixels on a screen or the sound vibrations of speech. Understood as a primordial phenomenon, language is not reducible to these physical phenomena but is what they bring to light – it is the very meanings or field-patterns of significance that they bring to light, the latter being essentially field-patterns of awareness composed of qualitative tones and intensities of awareness.

Fundamental Biology

Genetics, the foundation of modern neo-Darwinian biology, rests on a claim that we can 'decode' the 'book of life' by understanding its molecular constituents and structure. This is equivalent to claiming that we can understand a book by:

1. analysing the ink marks on its pages

2. identifying a basic 'alphabet' of these marks

3. cataloguing their combinations and permutations

4. seeking to explain how these permutations produce whole sentences, paragraphs and chapters — indeed the whole book.

A text is the two-dimensional surface of a multi-dimensional world of meaning. Molecular biology reduces the 'book of life' to its fleshly three-dimensional text. We do not understand a text because our brains 'decode' the ink marks on the page and manufacture a meaning from them. We understand the text, because as aware beings, we already *dwell within* the world of meaning it expresses. We can no more find scientific 'evidence' of the human being by medical research into the human body or brain than we can find evidence of human meaning in a text by chemical analysis of its ink and paper or mathematical analysis of the patterns of ink marks on the page.

The geneticist R.C. Lewontin has pointed out that 'the human genome' is itself a scientific myth. For each individual's DNA differs from others by around three million nucleotides. We tend to assume that Human Genomics is based on empirically proven, biological facts, whereas in fact, its basis are linguistic metaphors, which are then taken as literal 'facts' — in particular the linguistic metaphor of a 'book of life' without meaning and without an author. Talk of a 'book of life', of genetic 'instructions', that are 'read', of cellular 'communication', a molecular 'master language' etc. belong to the sphere of biosemiotics, the understanding of molecular structure as an autonomous sign system without any semantic dimension – a molecular text without any inner dimensions of meaning. The failure to acknowledge a semantic dimension to the linguistic metaphors employed in molecular biology leads to a number of basic fallacies in current biological interpretations of the medical significance of genetic research. For we can no more say that the 'causes' of disease lie in missing or 'misspelled' genetic sequences that we can say that the 'causes' of a poor text lie in its bad language or misspellings. And we can no more predict a person's behaviour or bodily functioning by analyzing their genes than we can predict what they will say by analyzing their alphabet or vocabulary. Just as words mean different things in different contexts so are genetic 'instructions' read differently by the body in different molecular, cellular, organic, natural and social environments. All that genetic biology can 'prove' is that if certain letters are missing then certain words cannot be spoken or may be physiologically 'mispronounced'.

A language is composed of a finite verbal alphabet and vocabulary but grants infinite potentials for the expression of meaning. The purpose of words is not to 'generate' only standard, well-formed sentences but to give form to potential meanings, in 'unstandard' or apparently 'abnormal' ways. Similarly, the human body is composed of a finite genetic 'alphabet' and 'vocabulary' but can give expression to infinite potentials and propensities of being.

The purpose of genes, as combinations of nucleotides comparable to words, is not only to allow the formation of 'well-formed' protein sentences and a well-formed fleshly text – a 'standard' or 'healthy' body or brain. It is to give form to unique individual potentials and propensities of being, unique inner field-patterns of significance, in however 'unstandard' or apparently 'abnormal' ways. From the standpoint of Human Genomics, a Down's syndrome child with extraordinary emotional sensitivity is biologically 'abnormal', whereas an adult political leader with a quite ordinary level of insensitivity to the bombing of civilians in a war is biologically 'normal'. From the standpoint of any Fundamental Biology any limitation on human genetic diversity, like any Orwellian type limitation on human linguistic diversity – the words people are allowed to use or create – will ultimately block the expression of human potentials. A race that cannot produce individuals *thought of* as 'mad' or 'handicapped' will produce no 'geniuses' either.

DNA is often compared to a molecular 'master language' informing the structure of all living organisms. This is a misleading comparison which reveals a basic misunderstanding of language. Organisms and their biological structures, like languages and their syntactic structures, are shaped by organizing field-patterns. These field-patterns, however, cannot be identified with their expression or representation in any one structure or set of structures, molecular or linguistic. To believe that one 'master molecule' in-forms all others is like believing that one master sentence, master text or linguistic meta-language can provide the key to understanding the structure of all other sentences, texts and languages – or that one poem or piece of music can exist or be created that expresses the essence of all others.

Biology as a science continues to ignore yet another fundamental distinction. between an organism and its environment as we humans perceive it with our own species-specific field-patterns of awareness, and the organism's own perceived environment, shaped by its own field-patterns of awareness. As the biologist Uexküll pointed out long ago, for a tick there is no such thing as a dog, sheep or human being in its environment — only mammals as such, which it perceives through their warmth. The entire relation between source fields of awareness and the manifest field-patterns that emerge from them can be compared to the relation between an ocean and the manifest life forms that arise within it. Just as any manifest or actualised field-pattern of awareness automatically patterns its own field of awareness, so do all the patterned life-forms within the ocean have their own uniquely patterned perception of their oceanic environment and the other life-forms within it. The way we perceive a shark, jellyfish or any other oceanic life-form bears no relation to the way the shark or jellyfish perceive each other and the other life-forms within the ocean. What any organism essentially is, is not a function of its bodily appearance, structure and behaviour as these appear to other organisms. An organism is essentially the embodiment of an organizing field-pattern of awareness, one whose manifest form will vary according to the field-patterns of awareness embodied in other organisms.

The field-pattern of awareness that constitutes the essence of any organism also patterns its own awareness, not only of other life forms but of the ocean — or any natural environment — as such. What the ocean as such — or any natural environment — essentially is, can also not be reduced to the way it is perceived by the life forms within it. Each of these life forms can experience itself in two different ways — as a consciousness separate and apart from other life forms and its perceived environment (our typically human mode of awareness) or as a part of that environment, connected to other life forms through it. Each life form however, is also a self-manifestation of its environment; different oceanic life forms, for example being a self-manifestation of the life of the ocean as such. What we as human beings experience as our own inner essence or self — and what constitutes the selfhood or essence of any oceanic life form, is the ocean's awareness of itself in the individualized form of a particular fish. The relation of environment to organism is one expression of the relationship between source fields of awareness and the manifest field-patterns of awareness or consciousnesses that arise from them — the source field knowing itself in and through the latter.

The organism as a physical phenomenon is its appearance within the patterned fields of awareness of other organisms. The organism as a primordial phenomenon — in its essence — is its own organizing field-pattern of awareness. This applies also to the human organism. Phenomenology has traditionally distinguished between the body of the human being as perceived from without, and the 'lived' body, the physical body as experienced from within. Fundamental or field-phenomenological biology understands the physical body as such as the outwardly perceived form of the human organism. The human organism itself is not simply the physical body as we are aware of it from within. It is a body of awareness. As such it unites two distinct fields of awareness. The first is the extensional field of our environmental awareness, embracing all the other bodies that we are aware of in that environment. The second is the intensional field of our bodily self-awareness. This inner-bodily self-awareness is not, in the first place, an inner awareness of our bodies which then colours our mood or subjective feeling tone. Rather it consists of patterned tones and intensities, densities and textures of awareness — 'feeling tones' for short. It is the latter that are themselves embodied in cell and organ tone, nerve and muscle tone, as well as being echoed in our tone of voice and in the overall 'tone' of our bodies. That is why we speak of someone being in sound health.

The word 'organism' derives from the Greek *organizein* — to play on a musical instrument or *organon*. The human organism is the instrument or *organon* with which we give extensional, bodily form to inner feeling tones, embodying them in nerve and muscle tone, cell and organ tone. It is also the instrument with which we translate inner field-patterns of awareness into bodily and behavioural patterns, into muscular and motor patterns, mental and emotional patterns, neurological and perceptual patterns.

The Eighth Fundamental Distinction

The Eighth Fundamental Distinction is between the human body, mind or soul on the one hand, and the human organism on the other. The body is the outwardly perceived form of the organism. The 'mind' is the internally represented environment of the human organism. The 'soul' is the aware interiority of the organism, composed of felt tones and textures, densities and intensities of awareness.

Fundamental Psychology

A Fundamental Psychology is also a Fundamental Physics, both having to do with the Fundamental Dynamics of resonance. To begin with we need to distinguish psychical, inergetic or intensional resonance on the one hand, and physical, energetic or extensional resonance on the other. The space of our inner psychical resonance with a piece of music or with another person is by no means the same as the extensional space within which this resonance manifests as a physical oscillation or vibration of molecules of air. The physical resonance may be a necessary condition for the psychical resonance, but it is certainly not a sufficient condition. Unwelcome music blaring away in an adjacent room or house may be loud enough to cause one's whole body to vibrate in physical resonance with it, but this is no guarantee that we are in resonance with it psychically. And when a composer creates, no physical sounds or resonances are even necessary as a condition of psychical resonance with those inner tonal field-patterns and intensities of awareness that then find expression in musical compositions and their physical, instrumental performances.

Psychical resonance in other words, is no less real or primary than measurable physical vibrations and their resonant effects. It is resonance as a primordial phenomenon, in contrast to resonance as a physical phenomenon. Having distinguished between psychical and physical resonance however, we must also recognize that resonance as such — Fundamental Resonance — is the fundamental link between the psychical and the physical, intensional and extensional reality, inergy and energy. Fundamental Resonance is morphic resonance understood in the fundamental sense, linking intentional, inergetic field-patterns of awareness or significance with their energetic expression in patterned fields of awareness – in experienced or observed phenomena.

Fundamental Psychology, understood also as Fundamental Physics and based on the Fundamental Concept of Morphic Resonance, is not something restricted to human psychology alone but an expression of the very nature of the physical universe or reality field. Fundamental Psychology is nevertheless fundamental to human psychology in all its aspects, and being so, should be and is able to provide a deeper, more fundamental account of the human psyche and human psychological capacities. Your very ability to read this paper, for example, is itself a prime example of morphic resonance understood in the fundamental sense – the patterned form of the words on the page being in resonance with certain underlying field-patterns of significance and thereby evoking a resonance with these patterns in the reader. The reader does not read the words as such so much as use their own wordless intuitive or felt sense of meaning to read their resonances and the patterns of significance they express. Felt sense or meaning is the reader's own medium of attunement to and resonance with these patterns.

Morphic resonance is also the basis for our understanding of the spoken word, whose phonic form conveys meaning through resonance with field-patterns of significance. Where we do not understand another person's words, this is because we ourselves have difficulty in resonating with the particular field-patterns of significance that they give form to, and/or because they themselves are not in full resonance with the field-patterns they seek to express. Perception too, is a prime example of morphic resonance. Infants learn to give perceptual form to objects as expressions of field-patterns of significance, even before significance is attached to these objects through naming. Non-verbal interaction and communication is also an expression of morphic resonance. When someone laughs, frowns or smiles, the very form of their facial expression amplifies their resonance with a particular field-pattern of significance, and tends to evoke a similar resonance in the beholder. A false smile, frown or laugh on the other hand, is defined precisely by a lack of inner resonance and thus a lacking power to induce resonance in others.

Habit is the fixation through self-resonance of a behaviourally expressed field-pattern. Posture, gestures and body language are the expression of self-resonant muscular and motoric patterns. Learning is an actualisation of new potential field-patterns of mental or motor activity, expressed as mental patterns and skills, language patterns and skills or motor patterns and skills. Once actualised, these patterns are stabilized through self-resonance. We do not need any more to think about what we are doing when we speak, walk or drive a car. We know what we are doing tacitly or implicitly – simply through attunement to an inner field-pattern of activity already stabilized through its actualisation in the learning process. It is not knowledge in the form of 'knowing about' but a direct knowing awareness attuned to potential field patterns of awareness and actualised in mental and motor patterns, bodily and behavioural patterns.

Philosophers of science and scientific psychologists have both developed elaborate theories of the possible relation between the human psyche and the brain. The fact that books can be published today with titles such as "How the brain thinks" is evidence of how little progress has been made. As Heidegger already pointed out, organs such as the brain possess functions but no capacities. A brain can no more 'think' than a pen can 'write'. As an organ, it is, like any other an organon or instrument, possessing no capacities of its own. Once again we come across a fundamental misconception that arises from attempting to reduce our very capacity for conscious thought and perception to some 'thing' – in this case the brain – that we can perceive and think about.

There is no doubt that the brain plays an important role in human thought and perception, sensation and emotion. But that is because neurological patterns and processes are themselves a manifestation of inner field-patterns of awareness and are in resonance with these patterns. The fact that chemically or electrically induced alterations to neurological patterning induce definite psychological effects such as mood changes is no proof whatsoever that conscious awareness is a product of the brain and its neurological patterning. Instead what we are witnessing is a disruption of a natural resonance between neurological patterning as such and those patterned tonalities and intensities of awareness that are experienced as moods and mood changes. That is why the 'efficacy' of most psycho-pharmaceutical drugs lies not in elevating mood or improving brain functioning but in suppressing it – replacing it with dull, catatonic moodlessness or a bland euphoria that is not so much a mood as a form of suppression of moods and moodedness.

Psychoanalysis can probe the meaning of dreams but only Fundamental Psychology can say what dreaming as such essentially is. Firstly, dreaming is a primary example of how an extensional spatial field of awareness can open up within an intensional or psychical field, one which in itself occupies no extensional space. Secondly, it is an example of how phenomena manifesting within a field of awareness – the figures that manifest in our dreams – are themselves figurations of that field – shapes or figurations of awareness. The dream field of awareness includes not only the experienced events and environment of the dream but the dreamer's own self-experience – our self-experience and our experience of particular events in the dream environment being in constant and dynamic relation. The very shape and texture of our bodily self-awareness changes in relation to our experience of different dream events or environment, a glorious or sinister dream landscape being part of the larger body of the dreamer, or a flying dream constituting a lived, bodily experience of a specific relation to that environment. Thirdly, dream events do not require some sort of interpretative decoding but speak for themselves. One reason they do so is that they are not just symbols of our felt experience of life, but are a part of our life experience — directly experienced events, whose meaning is itself directly felt. More so than in waking life, the meaning of dream experiences is itself something that is directly felt or experienced, that directly felt or experienced meaning is itself directly and immediately given form within the dream itself.

Dreaming is itself a direct experience of the dynamic relation between directly experienced meaning or felt sense and its symbolic expression in experienced events. It is also one of many ways in which our lived experiencing is itself a language, one in which each dream event speaks for itself. Thus the dreamer does not look for some hidden or 'unconscious' meaning in a dream in which they find themselves climbing a tall mountain and then poised on a precarious ledge. They need to follow what Freud himself did in practice, namely attend to the deeper resonances of the specific words and language in which a dream is recalled – for example by asking themselves what 'mountain' is it they are struggling to climb in life or in what way the find or feel themselves 'precariously poised', 'on a ledge' or 'on edge' in their life. Then not only will the dream have explained itself through the language of the 'rebus' (a visual word such a as 'mountain') it will also have reinterpreted the dreamer's waking life. The language of dreams in other words, defies any understanding of their symbolism as something to be interpreted in terms of other symbol or words that the ones they offer us. Fundamental Psychology distinguishes between experiencing as a process and its formed products. Thoughts and emotions, memories and mental images, localized bodily sensations or particular dream objects, are formed products of experience. These arise from, in-form and form part of the process of experiencing as such. As a field-phenomenological psychology, Fundamental Psychology understands the process of experiencing as a dynamic relation between an underlying field-state of unformulated awareness and its formed experiential products – the phenomena or contents of consciousness that manifest within this field. Unformulated awareness whilst pre-verbal, pre-reflective and pre-conceptual in nature is no more unconscious or undifferentiated awareness than a pre-verbal and pre-reflective comprehension of a piece of richly textured music.

Fundamental Psychology lays the basis for a fundamental paradigm shift in the focus of psychotherapy:

- from relations or associations between psychic contents within a field of awareness to psychic fields and field dynamics of awareness as such.

- from products of experience to the process of experiencing.

- from phenomena that a therapist or client is aware of to the underlying field-states of awareness from which they emerge.

- from thoughts and emotions to underlying moods or feeling tones — understood as qualitatively toned and textured field-states of awareness prior to both thought and emotion.

- from formed and formulated aspects of an individual's self-experience to the unformed and unformulated awareness from which they arise and which they give form to.

- from approaches which seek to make sense of an individual's feelings to a psychology of felt sense — felt meaning and intent.

Up till now psychotherapy has been dominated by a form of emotional reductionism — the attempt to reduce the meaning of a client's experience to specific feelings or to derive the meaning of their experience from those feelings. Feelings are treated as pre-given entities or 'internal objects' to be recognized or discovered, observed and objectified, 'made conscious' and communicated, experienced and expressed, 'worked on' or 'worked through'. Experienced feelings and cognitively grasped meanings are seen as intrinsically interrelated, but no acknowledgement is given to their common source or ground — to directly experienced meaning or 'felt sense'.

Fundamental psychology shifts the focus of psychotherapy from feelings 'and' their meanings to felt meaning as such, and to the way this is experienced and expressed by both therapist and client. Felt sense is not felt sensation but felt sense — felt meaning or intent. Gendlin's philosophy of felt sense acknowledges that meaning or sense is not only something verbally represented, signified, symbolized or somatised. It can also be directly experienced or felt — independently of its expression in words and propositions, symbols and concepts. Were this not the case it would be impossible for us to feel at a loss to express an experienced meaning or intent. Nor would we be able to assess the appropriateness or 'fittingness' of our words and concepts to that felt meaning or intent. We are most in touch with felt sense when we search for the right words or action to understand and respond to what touches us in our field of awareness.

Dimensions of Felt Sense:

- Words are understood through our wordless felt sense of meaning.

- Speech is the verbal articulation of felt sense.

- Listening is attunement to felt sense.

- Emotion is an intensification of felt sense.

- Action is guided and impelled by felt sense.

- Memories arise from the recall of felt sense.

- Concepts are encapsulations of felt sense.

- Bodily sensations are embodiments of felt sense.

- Comprehension is a widening grasp of felt sense.

- Dreaming is the experienced symbolization of felt sense.

- Objects are symbols of felt sense — they give perceptual form to felt sense in the same way that words give conceptual form to it.

- Experiencing is the dynamic process of giving form to felt sense, and in doing so both in-forming and trans-forming felt sense.

Not just words, but things too, are what they mean to us. Both the spoken word 'tree' and a perceived tree give determinate form to an otherwise indeterminate field of potential patterns of significance that we attune to through felt sense. These potential gestalts or patterns of significance can take form as verbal patterns, logical patterns, perceptual patterns or action patterns. Felt sense is also a felt resonance with field patterns of potential significance. A field-state of awareness is a field of potential patterns of significance which we cannot mentally identify as a determinate 'feeling' but which we nevertheless sense in a bodily way as a texture of felt tones, intensities and colourations of awareness.

Fundamental Science and Felt Sense

Like Rupert Sheldrakes's concept of Morphic Resonance, Eugene Gendlin's concept of directly cognised meaning or "felt sense" is a Fundamental Concept. Both contain within the seeds of one of the most significant and fundamental paradigm shifts ever to occur in our understanding of the nature of science and scientific knowledge, of the human being, the human body and human relations. Not only does this paradigm shift have immediate applications in the area of psychotherapy but profound and revolutionary implications for our understanding of health and medicine in general. These implications and applications can only be adequately articulated through the conceptual framework of Fundamental Science, which allows us to understand what exactly felt sense is, why attending to it brings benefits, and how it constitutes a form of Fundamental Cognition.

Understood from a field-phenomenological perspective, felt sense is a direct awareness of particular field-states of awareness. These are not composed of experienced phenomena which we happen to be aware of and can then attach some meaning to. Instead they are composed of qualitative tones and intensities of awareness, which, like musical tones, form themselves into meaningful patterns. Like moods, field-states of awareness are what tone and colour, shape and pattern our experience of ourselves and the world. At the same time however, they are not reducible to any actual phenomena we experience in ourselves or in the world. Felt sense is an attunement to different potential field-patterns of significance latent within those field-states.

Every field-state of awareness consists of different potential field-patterns of significance – intrinsically meaningful 'arrangements' of these tones of awareness. These field-patterns of significance arc the source, not only of language and its verbal patterns but of perceptual patterns – indeed of all experienced phenomena. Experienced phenomena are themselves patterns of events emerging from fields of awareness and giving form to specific field-patterns of significance. Both language and experience give expression to these field patterns. From a field-phenomenological perspective, words do not merely represent or describe 'things' – experienced phenomena whether physical or psychical. Rather things themselves, whether physical or psychic objects, are intrinsically meaningful – serving as words or signifiers in our own experiential language, and providing its basic vocabulary. We do not depend on language to make sense of experience. Instead experienced phenomena possess an intrinsic sense that we feel directly.

It is of the utmost importance to recognize that felt sense is an attunement to those inner field-patterns of significance that find expression in both language and experience. Both words and things, language and the experienced phenomena it describes, give outer form to these patterns, and in doing so function as signifiers of these patterns. The power of these signifiers is such that we no sooner have a felt sense of a particular field-state of awareness – a set of potential patterns of significance – than we translate it into signifiers, into formed percepts or concepts, mental images or words. The latter both belong to the realm of phenomenal signifiers – serving to give phenomenal form to felt sense. But any phenomenal signifiers – any pattern of words or events, observations or experiences, can give form to only one potential field-pattern of significance. In doing so it leaves other patterns unmanifest – unperceived, unspoken, unthought. Were it not for felt sense, these other patterns would remain also unsensed and unfelt, for felt sense is the 'excess' of meaning that no phenomenal signifier can capture or contain.

Signifiers of any type, whether verbal or experiential, not only give expression to felt sense but also enframe it in a particular way, giving form to particular field-patterns of significance and leaving others unexpressed. The result, unfortunately, is an entire culture and civilization in which meaning as such is identified not with intrinsically meaningful dimensions of the world, but only with formal signifiers that 'refer' to things in the world i.e. words or visual signs, mental images or mathematical signs, scientific diagrams or observational 'data', religious symbols and scriptures etc. Because of this, meaning is seen primarily as a function of language and sign-systems. Experience is either placed in a secondary role as that which is 'signified' by these sign-systems, or else experienced phenomena are seen as a sign-systems in themselves – but as closed systems in which the meaning of any sign is purely and simply a function of its relation to other signs within the system, and not to any inner patterns of significance transcending that system ('post-modernism' and 'biosemiotics').

Our felt sense of the excess and unmanifest significance of experienced phenomena — objects or events, people or situations, words — is treated as something merely 'subjective', or as some dubious 'sixth sense' belonging to the realm of the extraordinary and paranormal. And in a certain sense this is right, for felt sense is a sixth sense, but this sixth sense is in fact the basis of all the other senses, whose function is to give perceptual form to those inner field-patterns of significance which we first attune to only through felt sense itself.

Gendlin points out that people tend to refer back to their unformulated felt sense of particular dimensions of their experience only when they are at a loss for words – when they lack fully-formed signifiers that can express certain unformulated dimensions of their experience itself. The reason why attending directly to felt sense helps us find words or other signifiers that express what we mean is that through felt sense we attune to the inner field-patterns of significance that are the very basis of both verbal patterns, mental images, and conceptual structures. The importance of talking specifically of signifiers of felt sense is that they include more than just words. A situation, symptom, emotion or life event is also a signifier, one whose meaning we can not only give form to in language but also feel directly in a bodily way. The relation between language, experience and felt sense can be visualized in the form of an inverted triangle.

Diagram 1

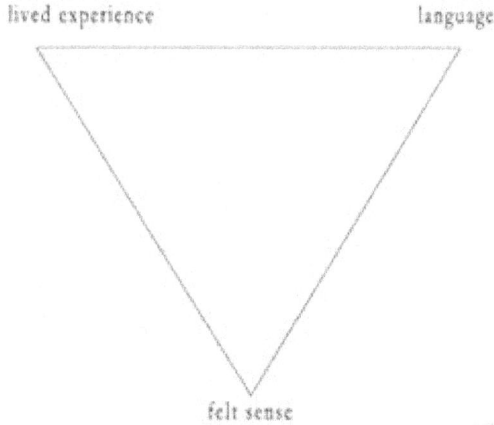

What I call phenomenal signifiers are represented by the entire line connecting language and experience, and include all dimensions of formed or formulated experience from identifiable events and emotions to the words we use to describe and name them. People do have a felt sense of the significance of experienced events and situations as physical signifiers just as they also have a felt sense of the meaning of words and do not identity this with their given meaning or denotation as verbal signifiers. But like the linguist and semioticist they tend to see language as a sign system whose main role is to refer to, represent or 'signify' experience (their own or others). They ignore its other, and most crucial role – that of giving form to felt sense and the unformulated dimensions of experience it links us to. But as the poet knows well, language can only truly do justice to our experience of the world if it is in resonance with felt sense – if it helps to give form to the felt, but unformulated or unmanifest significance of particular experiences. If the lines of the triangle linking language with experience on the one hand, and felt sense on the other are not distinguished – not only theoretically and experientially – the consequences are significant. Either the expression of felt sense is sacrificed to the literal signification of language, enframed and eclipsed by conventional social and scientific languages, or else language becomes solely a tool of the imagination, used to express felt senses unrelated to actual experiences and shared social realities.

The Dyadic Field

The self-defined role of psychotherapists and counsellors is to help the client to find significance in their lived experience and to give form to unformulated dimensions of this experience. But in psychotherapy too, felt sense is easily enframed and eclipsed by language. For a therapist may no sooner gain a felt 'empathic' sense of something important to the client than this felt sense is translated into a verbally communicable emotion or insight – in other words into cognitive or emotional signifiers. The original, unsignified and unformulated sense that was the source of these signifiers is then eclipsed by its translation, for it is now identified with and enframed by these signifiers. The result is that the therapist is no longer able to refer back to felt sense – to feel whether their 'interpretation' (cognitive and emotional) of a client's words or experiences is in resonance with the felt sense that gave rise to them.

Unfortunately the entire framework of psychotherapy and counselling training encourages this premature foreclosure and enframement of felt sense, guided as it is by the belief that listening to a client is merely a prelude to providing some form of verbal response. Truly deep listening on the other hand is not simply a prelude to some form of outward response to another person. It is the capacity to attend and attune to felt sense and stay with it long enough to find signifiers that are fully in resonance with it. A fully resonant signifier will not enframe or eclipse felt sense but amplify it. The capacity to stay with felt sense is not only the condition for transforming felt sense into resonant signifiers. It is also the condition for transforming it into a felt resonance with others – with their own felt sense of the inner meaning or significance of their words and mental images, emotions and life experiences. Felt sense is the gateway to a felt resonance with the field-states of awareness and field-patterns of significance that find expression in another person's words and experiences. Deep listening is therefore resonant listening in this double sense – the transformation of felt sense not only into resonant signifiers, but also into direct field-resonance with another person. Both these forms of resonance amplify felt sense and amplify one another.

From the point of view of any Fundamental Therapy, the theoretical models on which psychotherapy and counselling are currently based are a substitute for a deeper understanding of listening itself as felt resonance, and the therapeutic and counselling relationships are a substitute for deep, resonant listening in society at large – and in all the client's other life relationships. Just as somatic medicine is based on a biological reductionism – reducing a patient's felt dis-ease to a diagnosable disease, so is psychotherapy governed by a tendency to emotional reductionism – identifying meaning with conscious or unconscious emotions. This emotional reductionism is but the flip side of a culture which identifies meaning with intellectual concepts. But it is also the expression of a more basic cultural framework – reinforced in many forms of therapy training – which identifies felt sense or meaning in general with its formed expressions in phenomenal signifiers such as words and body language, existential facts or subjective feelings. Paradoxically, the very attempt on the part of physicians, psychotherapists or psychoanalysts to find out what a client's problems are 'really' all about then turns into an attempt to reduce felt sense to one 'central' or 'basic' signifier among others.

Gendlin's philosophy of felt sense is therefore of fundamental significance for the future of both psychotherapy and somatic medicine, and above all, for the training of therapists of all sorts. It provides the missing link between psychotherapy, on the one hand and somatic medicine. That is because felt sense or meaning is indeed something only experienced in a bodily way – but precisely for that reason it is easily confused with bodily sensations and symptoms, emotions or 'energies'. What Fundamental Semiotics adds is a deeper understanding of the relational dimension of felt sense as a gateway, not only to the felt self but to a direct felt resonance with others.

The failure to distinguish felt sense from phenomenal signifiers is reflected in a failure to fully acknowledge a whole range of parallel distinctions of profound scientific as well as spiritual significance – a failure not only to signify these distinctions in language but even to experience them in everyday life. Medicine totally fails to acknowledge the distinction between a patient's felt dis-ease and diagnostically-labelled diseases or disorders – with the result that patients are themselves encouraged to confuse or identify the one with the other. And neither semiotics nor linguistics have any place for felt sense or the unformulated dimensions of experience it links us to, concentrating only on sign systems, formal signifiers or the formulated dimensions of experience and phenomenal signifiers they refer to.

Fundamental Medicine

From the point of view of Fundamental Medicine, 'health' is not normal 'functioning' nor even 'well-being', and nor is ill-health simply dysfunction or lack of well-being. Fundamental health is 'value fulfilment' – a capacity to fulfil our innermost values or potentials of being. Ill-health is a natural and meaningful part of the health process by which we learn to express and embody these potentials of being or spiritual genes. Illness is a gestation process, a form of spiritual pregnancy by which, through a more or less painful or complicated 'labour', we are led to acknowledge and give birth to new aspects of ourselves. Diseases of the body and mind are the expression of a felt dis-ease of the inner human being. When we are ill we do not 'feel ourselves'. That is not because we are the victims of foreign bodies such as micro-organisms, toxins or antigens – what immunology and oncology describe as 'non-self' molecules or cells. Fundamental dis-ease, our 'not feeling ourselves', arises because we are on the way to 'feeling another self' – but have not yet found the way to embody and express that new sense of self.

What I call 'self-states' are field-states of awareness with which we are completely identified. These are experienced as tonalities and textures of awareness, which, like moods, so completely permeate our felt sense of self and so colour our awareness of the world that we are hardly aware of them – unless they change. New field-states of awareness however, may be initially experienced as threatening – in dissonance with an established sense of self, or in dissonance with mental and emotional patterns that express that existing sense of self.

What we call the 'self' has a field character, consisting of a unique range of self-states – uniquely toned intensities and colourations of awareness, bearing within them unique field-patterns of awareness that shape our personal reality. Illness, disease and dissonance are a natural part of the health process by which we learn to resonate with new aspects of our own larger identity or self-field. Illness is the bodily expression of dissonant organismic field-states of awareness in which our beliefs and world picture distort our experience of new self-states and deny expression to them. They form part of a mental immune system whose function is not to protect our bodies from infection or colonisation by foreign bodies – so-called 'non-self' cells – but to protect our self-experience from infection by aspects of ourselves that we still experience as foreign or 'non-self'.

Our self-experience not only shapes and colours our experience of other people and the world, but is also reshaped and recoloured by it, allowing us to experience new aspects of ourselves, to identify with new self-states or selves. In this way we grow as individuals, learning to give expression to new potentials latent within our own self-field, doing so through resonance with the self-fields of others and with their actual patterns of self-expression. Health, from this point of view is not a continuous state of undifferentiated well-being, balance or harmony but a continuous process of change or metamorphosis, based on a balance of balance and imbalance, a harmony of resonance and dissonance through which we give expression to new psychic potentials of our inner being – our as yet unexpressed psychic genes.

Through respiration, circulation and metabolism, our bodies have no natural difficulty in absorbing foreign matter and reconstituting themselves from it. Our own inability to process the raw material of our experience, using it not only to reconstitute, but to recreate and reshape our selves from it – hinders and interferes with the organismic field-patterns that shape and sustain our bodies' own respiratory, circulatory and metabolic patterns. Fundamental Medicine understands physiological disorders as a precise symbolic expression of disorders of psychic respiration, circulation, metabolism and immune functioning. These in turn are the expression of a felt dis-ease which is the expression of dissonant field-states of awareness – a lack of resonance with new and hitherto foreign field-state of awareness that form part of our larger identity or self-field. This lack of resonance with aspects of ourselves may be experienced initially as dissonances in our relationships with other people or our environment, as 'stress' induced by dissonances we experience in other people and our environment, or as 'cognitive dissonance' – beliefs that are in conflict with each other or with our own internally sensed environment. What Eugene Gendlin calls "directly experienced meaning" or "felt sense" is the unformed or unformulated, pre-verbal and pre-symbolic dimension of experience that constitutes the very essence of organismic awareness.

The human organism and our own organismic awareness play an important role in the health process and the way in which new and unfamiliar field-states of awareness and a new and unfamiliar sense of self may manifest in illness. Organismic awareness is not body awareness. Body awareness is awareness of particular bodily sensations or symptoms. Organismic awareness does not consist of any sensations of symptoms we are aware of in our bodies but rather in the felt tone and texture of our bodily self-awareness as such. Fundamental dis-ease is first and foremost a muddied or dissonant field-state of bodily self-awareness. The individual may experience this own felt, organismic sense of dis-ease, however, only through felt bodily sensations or symptoms of a particular sort.

The Ninth Fundamental Distinction

Another distinction central to Fundamental Science is one of fundamental significance in human psychology and physiology, and of fundamental relevance to psychotherapy and somatic medicine. That is the distinction between both intensional and extensional field-patterns on the one hand and field-states on the other. Fundamental Science defines a field-state as a dynamic relation between an intensional or potential field-pattern of awareness and its actualisation. This relation can take the form of resonance or dissonance. Resonance stabilizes and sustains the actualisation or physical manifestation of a potential field-pattern. Dissonance, on the other hand can distort that manifestation, degrade a previously resonant manifestation of it, or deny expression to a previously unmanifest pattern. Resonance and dissonance go hand in hand. Without dissonance, dominant field-patterns cannot distort or degrade in a way that allows new field-patterns to manifest. At the same time dissonance can also distort or deny expression to those new field-patterns.

Organismic field states are stabilized through resonance between inner field-patterns of awareness and their manifestation as mental and emotional patterns, sensory and motor patterns, physiological and neurological patterns. Lack of resonance can result in felt dis-ease i.e. a muddied, hollow or discordant field-state of bodily self-awareness – an 'unsound' state lacking a clear resonant tonality. That is why we speak of people 'sounding well' or being in 'sound' condition. Dis-ease can also arise from resonances or discordances between individual field-states, field-patterns and field-tonalities of awareness and those manifest in a person's familial and social fields. When someone speaks of feeling socially 'stifled' or of having no 'room to breathe' this is not usually meant in a literal, bodily sense, but nor is it a 'mere' bodily metaphor. It is a description of a felt organismic state. Conversely, however if someone breathes more freely as a result of feeling their 'spirits' lift, then it is their actual bodily breathing that is the metaphor – a living, biological 'metaphor' of their inner being. A person can jog or exercise, or practice Yogic breathing exercises for hours, days or years without it significantly affecting their fundamental respiration – without it bringing new sources of spiritual meaning and inspiration into their lives. But a person can be neither spiritually inspired nor dispirited without it being instantaneously embodied in their physical breathing. The organism is the instrument with which we constantly translate states of being into mental and physical states, and transform basic capacities of our being into organic functions.

According to Martin Heidegger "We cannot say that the organ has capacities, but must say that the capacity has organs....capability, articulating itself into capacities for creating organs characterizes the organism as such."

Respiration, for example, is not merely an organic bodily function but the embodiment of a fundamental capacity of our being. That is the capacity to engage in a rhythmic exchange with the 'atmosphere' of our life-world – 'breathing in' our own awareness of it, drawing meaning and inspiration from it, and in turn allowing our awareness to flow back out into it – whether as a simple exhalation of breath or as meaningfully shaped and toned exhalation, as speech. At what point does the air we inhale become a part of us? At what point does our exhaled air cease, not only to be a part of our bodies but a part of us? When we draw into our awareness a 'breathtaking' landscape or an 'idea', we feel moved to inhale and then exhale deeply. Why? Because breathing is the embodiment of our fundamental organismic capacity to fully take into ourselves our awareness of something other than self, and in turn allow that awareness to flow out again into the atmosphere or field of awareness linking us with the world. The words 'respiration', 'inspiration', 'aspiration' etc. come from the Latin *spirare* – to breathe – just as the Greek word *psyche* originally meant the 'breath' that vitalised an otherwise lifeless corpse (*soma*). To speak in a modern way of the 'psychosomatic' dimension of breathing disorders such as asthma, to either claim or dispute their 'psychogenic' causation therefore misses the point. It ignores the question of what breathing as such fundamentally is – not as an organic function of our body but as an organismic capacity of our being. Changes in the pattern and flow of our bodily breathing embody differently patterned flows of awareness.

Specific organic dysfunctions such as respiratory, circulatory, digestive dysfunction are the manifestation of the relation between inner organismic capacities and their embodiment in organic functions – for example the relation between an individual's inner or psychical respiration or metabolism and their physical respiration or metabolism. Outer metabolism is the functioning of the body in digesting and metabolising foodstuffs. Inner metabolism is the individual's capacity to digest and metabolise their own experience of themselves and the world. Every experience of the self is an experience of something or someone other than self – whether another person, a piece of music, or a percept of any form. Conversely, every experience of something or someone other than self affects our self-experience. It is not the same self we experience washing up, being with a close friend or partner, engaging with another person professionally or participating in a social or mass event.

Interaction with the world and other people is a way of expanding our identity of felt sense or self-experience by incorporating elements previously perceived as 'other than self'.

What I call the mental immune system (MIS) governs the relation between our experience of others and otherness and our self-experience, maintaining a more or less rigid or flexible, closed or permeable boundary between that which we experience as 'self' and that which we experience as 'not-self'. Paradoxically however, over-active inner or mental immune defences can stretch and ultimately weaken the body's own immune functioning. By shutting down, not 'letting things get to us' we create a situation in which it is our bodies that do the 'letting in' – becoming biologically vulnerable to antigens or 'non-self' elements such as viruses. The illness we contract, however, may in turn have a meaning, allowing us to relax our inner immune system and mental defences, giving us time to catch up with ourselves and to process or digest particular experiences. Just as the body incorporates foreign elements whilst maintaining its patterned integrity so does the self. But whilst the body stops growing, the self never does – needing not only to maintain a sense of identity but to expand that identity. It does so by accepting new field-states of awareness as self-states, new ways of experiencing the self and the world coloured by a specific feeling tone that was hitherto felt as unfamiliar or foreign.

Psychic and somatic dimensions of dis-ease are by nature distinct but inseparable, the human organism itself being a psychic body of awareness whose field-patterns are the foundation of physiological functioning. Dis-ease is a field-state of organismic or bodily self-awareness expressed in somatic and emotional states. However it is of fundamental importance to recognize that the latter are themselves meaningful signifiers of a felt sense of dis-ease which is not intrinsically pathological, but the harbinger of a new sense of self – part of the health process. Somatic and emotional states are themselves somatic and emotional interpretations of a felt organismic dis-ease. Like interpretative medical diagnoses themselves, however, they play an important role in giving that dis-ease a pathological character.

Conventional medicine, not only orthodox medicine but also many forms of alternative medicine, sees symptoms as literal signs of an underlying somatic or psychological pathology rather than as metaphorical signifiers of a felt sense of dis-ease. The physician's first act is to separate the patient as a human being from their symptoms, to objectify the latter and to reduce them to signs of some 'thing' lying behind them. The therapeutic relationship takes the form of a 'We and It' relationship. Diagnosis is based both on the patient's verbal reports and on the results of examinations or tests. Both the patient's words or verbal signifiers and symptoms or somatic signifiers they describe are treated as signs pointing to an underlying disorder or disease which constitutes their 'cause'. This is rather like looking for the physical 'causes' of a person's words or body language rather than understanding their meaning. The human body is the fleshly three-dimensional text whose inner dimensions of meaning cannot be discovered through any internal physical examination or testing. Both the patient's words and the symptoms they describe are symbols or signifiers of a felt sense of dis-ease with many layers of meaning.

As Foucault wrote: "To ask what is the essence of a disease is like asking what is the nature of the essence of a word." Our felt understanding of the sense or meaning of a word always has to do with connotations that transcend its given meaning or denotation. Just as the same words can have a different felt meaning to different people, so can the same disease symptoms. This felt meaning may not however be manifest, visible, or expressible. It belongs to the realm of unformulated experience.

X-ray photographs, CT or magnetic resonance scans and thermal imaging all show a different picture of the human body and brain. None of them reveal the organizing field-patterns of awareness that constitute the human organism, and the manifold fields it inhabits. Nor do they show the role of an individual's thoughts and beliefs in shaping the physiological and physiognomic, bodily and behavioural expression of different organismic field-patterns. Whilst it is true that cognitive behavioural therapy, and alternative medicine and the new field of psycho-neuro-immunology all offer accounts of how a person's beliefs can affect their bodily well-being, forging evidential or speculative links between 'mind' and 'body', none of these accounts distinguish the human 'mind' or 'body' on the one hand from the human organism on the other, or recognize mental and physical states as the expression or embodiment of organismic states – field-states of awareness.

A field-state is a dynamic relation between a potential field-pattern of awareness and its actualisation. This relation can take the form of resonance or dissonance. Resonance stabilizes and sustains the actualisation or physical manifestation of a potential field-pattern. Dissonance, on the other hand can distort that manifestation, degrade a previously resonant manifestation of it, or deny expression to a previously unmanifest pattern. Resonance and dissonance go hand in hand. Without dissonance, dominant field-patterns cannot distort or degrade in a way that allows new field-patterns to manifest. At the same time dissonance can also distort or deny expression to those new field-patterns. Dis-ease is the experience of dissonant, degrading and/or newly emerging field-states of the human organism. But what we call the 'mind' is an integral part of the human organism, a body of thought-patterns or beliefs, more or less in resonance with one another, which actively in-form these field-states or distort them in line with social thought patterns and medical belief systems. These constitute a type of social text inscribed into the very texture of the patient's experience of dis-ease whilst at the same time ignoring its social and relational context.

A secretary, who feels abused and humiliated by her boss but incapable of facing up to him for fear of losing her job, develops instead an 'angry' skin rash on her face. The GP or alternative practitioner she visits may know nothing of this situation and ask no questions that bring it to light. A diagnosis is made and treatment given which may or may not be effective. The greater danger, as in Zigmond's case study, is that the treatment is effective, for this may leave the patient with no option but to manifest or 'somatise' her dis-ease in another, perhaps more serious or life-threatening way. Fundamental healing is 'self-healing' in the deepest sense – discovering that hitherto unknown, unmanifest and unfelt self that is capable of meeting the challenges the individual faces.

Most forms of medicine seek to affect the human organism indirectly, from without. Mesmer practiced a form of direct organismic healing based on the resonant contact and communication between the organism of the healer and that of the patient, one well summed up by his follower Tardy de Montravel:

"The nerves of the two human beings can be compared to chords of two musical instruments placed in the greatest possible harmony and union. When the chord is played on one instrument, a corresponding chord is created by resonance in the other instrument."

Just as there are organismic counterparts to every organ and physiological function (in this case the skin and auto-immune function) so there are also inner psychical counterparts to physical phenomena such as space and time, closeness and distance, warmth and coolness, light and darkness, lightness and heaviness, sound and density, charge and polarity, electricity and magnetism. This applies also to the elements such as fluidity (water), solidity (earth) and gaseousness (air). The organism is composed of different combinations of these inner energies and elements – qualities of inner warmth and inner light for example. But it is of the utmost importance not to confuse our felt sense of these qualities with bodily sensations of some physical or vital 'energy'. The warmth we feel radiating from a human being for example, is neither a measurable property of their physical body (temperature) nor some mysterious inner energy that we happen to be aware of. It is a quality of their awareness of themselves and the world that others then feel bathed in and warmed by. Just as we can feel inwardly close to someone even though they are miles away, so can our feelings towards someone have a warm quality even though our bodies are freezing. Just as our organism, as a body of awareness, is not less but more fundamentally real than any physical phenomena we are aware of, so are inner closeness or distance, warmth or coolness, lightness and darkness, fluidity or solidity etc. more fundamentally real than their physical counterparts.

When we feel a person's warmth or see the radiance of their gaze we are not speaking of any physical heat or light emanated by their bodies. Nor are we merely speaking metaphorically. To believe so is to imply that the warmth or light we feel emanating from a human being is somehow less real than the measurable temperature of the human body or the measurable light reflected by their eyes. To talk, as many alternative practitioners do, of a person's bodily 'energy' and of 'energy medicine' seems to imbue our felt, organismic sense of other human beings with a more tangible 'objective' reality than a mere 'subjective' feeling about them. At the same time however, it is an evasion of the basic question of what is more real or fundamental – the human body or the human being, energetic relationships between bodies in space and time, or inner relationships between beings. To say that it is energy that links or relates things and people is one thing. Put the other way round, we can say that the essence of energy is relationality as such.

The principal instrument of Fundamental Therapy is the therapist's own organism and organismic awareness. The Fundamental Therapist is one whose own organism has become a new inner organ of perception, transforming their otherwise undifferentiated organismic awareness – 'felt sense' or '6th sense' – into a highly differentiated set of inner senses. These allow a direct feeling cognition of the inner field-tonalities, field-qualities and field-textures of another person's organism. The medium of this feeling cognition is 'feeling tone'. For like the audible tones and chords of music, inner feeling tones possess potential qualities of warmth and coolness, lightness and darkness, levity and gravity etc just as they also possess different tonal 'colours', different 'timbres', different elemental textures such as fluidity and solidity, airiness or fieriness and different sonic 'shapes' such as roundedness or angularity.

When we say of someone that they are 'warm and friendly', 'cold and hostile', 'remote and arrogant', 'heavy with remorse' etc. we use phrases which, as verbal signifiers, unite two quite distinct levels of meaning or signification, possessing both an organismic signification and a mental-emotional one. Let us say that someone with an emotional history of abandonment, abuse or deprivation of human warmth develops a 'cold' outer bearing. Others may perceive this bearing as emotionally hostile, defensive, aggressive, remote, cut off, arrogant. Indeed simply to describe someone as 'cold' has an emotional connotation. This shows that in language, as in life, people have not yet developed the capacity to distinguish fundamental qualities of organismic feeling tone such as inner warmth and coldness, light and darkness etc. from the way these qualities are emotionally experienced and expressed – both in others and in ourselves. But it is precisely the capacity to distinguish the mental-emotional and organismic significance of particular somatic or behavioural symptoms that is the foundation of Fundamental Medicine, Diagnosis and Therapy.

From the point of view of Fundamental Medicine it makes no more sense to regard sickness as an 'unnatural' deviation from health than it does to regard dreaming as an unnatural disruption of sleep, or nightmares as an unhealthy type of dream. The medical model of illness, based on the premise that illness is a meaningless deviation from health, is as outdated as certain pre-Freudian 'scientific' beliefs that dreams are meaningless discharges of neurological energy. Fundamental Medicine understands dreaming as one of two principal functions of the human organism. The other function is bodying – the organism being the instrument with which we embody our capacities of being and give bodily expression to feelings – to inner field-states, field-qualities and field-patterns of awareness. Bodying is not the same thing as 'somatising' – the unconscious production of somatic symptoms as an indirect expression and communication of psychic states. On the contrary, somatisation can be understood as an incapacity to body a felt dis-ease – to communicate it in a direct bodily way. In the contemporary psychoanalytic model of somatic disorders, somatisation is understood as a failure of symbolization. More specifically, it is seen as an expression of alexithymia – a deficient lexicon of verbal symbols by which to identify and process feelings. But ability to express and communicate feelings in words is inseparable from the ability to experience and communicate them in a bodily way. Psychoanalysis focuses only on the relative poverty or richness of a patient's verbal-emotional language or 'literacy', not on the poverty or richness of their emotional body language and its expressive vocabulary or 'lexicon'.

What I term 'bodying', as opposed to 'somatising', is in essence nothing more than engaging in this process of transforming formal and phenomenal signifiers into such isomorphic physiognomic signifiers. All serve to reconnect the patient with their own organism or 'dreambody', and automatically bring about what Gendlin calls a 'felt shift' in the patient's mental, emotional or somatic state. In terms of Fundamental Medicine, this felt shift is a shift in the patient's felt sense of self – a transition to a new self-state facilitated by the patient actively identifying with their own felt sense of dis-ease.

Today we speak of modern and traditional medicine, Western and Eastern medicine, conventional and complementary medicine, orthodox and alternative medicine, herbal medicine, 'energy medicine' etc. What I call Fundamental Medicine, however is not essentially another approach to medicine or a new form of medicine. It is a new and more fundamental understanding of what medicine as such essentially is. Only out of a more fundamental understanding of the nature of medicine as such can new forms of medical practice be founded which are grounded in this essence.

Central to all contemporary forms of medicine is a failure to acknowledge any intrinsic significance to the suffering that accompanies ill-health. Illness is identified with suffering and suffering with passive victimhood. The traditional aim of medicine has always been the alleviation or elimination of the suffering experienced by the sick person, and with it the diminution of any intense feelings, fears or anxieties accompanying or amplified by their illness. At the heart of almost all forms of modern medicine is the assumption that illness and suffering are not only undesirable in themselves but lacking in any intrinsic significance. Many thinkers have questioned this assumption. Balint argued that "Patients turn their problems into illnesses, and...the physician's job is to turn them back into problems". Groddek saw the meaning of illness as a warning sign: "Do not continue living as you intend to". Others have gone on to argue that not only is illness a warning sign but that successful 'treatment' or 'cure' is equivalent to simply disabling or destroying the warning light itself.

The fear of finding an inner symbolic significance to a patient's symptoms is that this is tantamount to 'blaming the patient for the illness'. But behind this accusation, so often levelled against those who question the accepted wisdom of the medical establishment, lies one of the most concealed and yet basic assumption of medicine in all its forms – the assumption that illness and suffering are something blameworthy and therefore 'bad' in the first place, something for which a cause or scapegoat must be found. To this accusation, Illich counters that "Health and suffering as experienced sensations are phenomena that distinguish men from beasts. Only storybook lions are said to suffer..." (Illich).

He reminds us that in the past, spiritual meaning was attached to illness, and different cultures each had their own rituals for the vital expression and communication of human suffering. Today's culture, on the other hand, regards the medicalisation and medication of suffering as the only rational response to it, and perceives the rejection of medical help as vain masochism. "Blaming the patient" means making them responsible for their suffering. Modern medicine, according to Illich, does the opposite. By depriving patients of responsibility for their suffering it deprives them of their own power to actively respond to it – to experience suffering itself not as passive victimhood but as responsible activity.

Children tend to actively express and embody their moods, their sense of ease or dis-ease, communicating it through their bodily countenance and demeanour. Parents however, often react harshly to any attempt on the part of the child to actively communicate their suffering through their bodily demeanour – telling the child, for example, not to sulk or brood, not to make a long face. The child is taught that suffering is something to be privatised and masked or else communicated only in a verbal way – that it must on no account be actively bodied. But as Heidegger intuited: "Every feeling is an embodiment attuned in this or that way, a mood that embodies in this or that way." (Heidegger). To body a feeling, even of dis-ease, is to let it be what it essentially is – "an embodiment attuned in this or that way, a mood that embodies in this or that way".

The term 'pathosophy' means 'the wisdom of suffering' or 'the wisdom of feeling' (pathein). It was coined by Viktor von Weizsäcker, who understood illness as an expression of the 'pathic'. He defined the pathic as "the essential suffering of a person that is related to that which they lack and that towards which they are aiming." The practice of modern allopathic medicine, on the other hand, is aimed solely at the medicalisation and medication of felt dis-ease, which is reduced to some form of diagnosable mental or physical 'pathology' and not in any way seen as a meaningful expression or embodiment of the pathic. Its dominant metaphor is that illness is a form of aggression or attack by foreign bodies or pathogens, and healing as war, a military campaign waged by the body itself and supported with medical interventions. Fundamental Medicine is founded on the basic metaphor of illness as pregnancy and healing as a form of midwifery or maieusis. It distinguishes between passively 'suffering' feelings and actively bearing and bodying them. It understands the fundamental role of the physician as that of midwife – someone who is able to bear with others in pregnant silence, to help them actively bear and body their suffering and in doing so give birth to a new sense of self and a new, more fulfilling inner bearing towards life.

The Tenth Fundamental Distinction

Scientific research aims at the accurate representation of relationships between things and between people. Ethics, on the other hand, is concerned with our relationship to things and to people. Fundamental Science is ethical in its very essence, for it is based on the recognition that our understanding of relationships between experienced or observed phenomena, indeed our very experience or observations of those phenomena, is an expression of our relationship to them. A deeper understanding of the manifest external relationships between things, like a deeper understanding of the manifest, external relationships between people, can only come about through a deepening of our own inner relationship to them, and with it an understanding of their own inner relatedness. The tenth distinction essential to Fundamental Science is between outer and inner relatedness as such. External relationships between phenomena in an extensional field of actualisation are their internal relationships to each other within the intensional source fields of awareness from which they emerge. Every experienced or observed phenomenon manifests within an already patterned field of awareness – its field of actualisation – possesses a definite external relation within this field. But emerging as they do from a non-extensional source field of awareness, phenomena are also inwardly linked to one another by virtue of being common manifestations of this source field — patterned forms or figurations of awareness, linked to one another by a common field of awareness.

Fundamental Topology

So far no adequate topology exists that embraces the relationship between extensional and intensional fields and conceptualises the nature of non-extensional or intensional space. Yet a relational field topology of this sort is fundamental to any deeper understanding of the nature of bodies and bodyhood as such. Fundamental Topology is relational field topology. It is founded on the recognition that any three-dimensional body is nothing more nor less than the surface boundary, membrane or interface between two fields, one field enveloped by that boundary, the other enveloping or environing it. This being the case, we are immediately confronted with a fundamental distinction between two types of space: space and counter-space.

The diagram below appears to show a black circle against a white background. But we could equally well interpret it as a white circle with a black interior. The figure as such – the circle – is neither black nor white, foreground or background, but the boundary between the black field that it envelops and the enveloping white field around it. It is only a bounded black 'foreground' figure from an ordinary spatial point of view. From a counter-spatial perspective the figure is an internal configuration of its enveloping white field.

Diagram 2

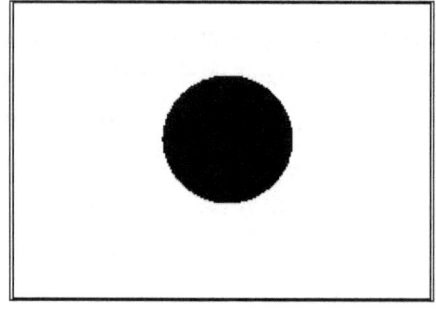

From our ordinary spatial perspective, which focuses on foreground elements in our visual field the circular boundary of the disc *appears* to bound the black disc, i.e. contain the black region or field it surrounds. That is why, most people, when asked what colour the disc is, will immediately say black. Yet as we have seen, from a counter-spatial point of view however, the circular boundary that gives form to the disc is an internal boundary of the enveloping white field around it. Given that a disc is by definition circular, could therefore be said that the disc itself is white, being not simply a foreground figure but an internal configuration of its own white background field. The argument can be simplified if, instead of a black disc we just picture a circle drawn in black or any other colour on a white background. For it is then clear that the 'circle' is just as much defined constituted by the white areas within and around it as by any colour we use to draw it.

If the topology of a basic two-dimensional figure such as a circle is grasped in this way, it becomes clear that we cannot, as we tend to do, identify are particular shape or figuration with whatever it *surrounds* but must see it also as an internal figuration of what surrounds it – of its own surrounding field. This applies to three-dimensional forms too. Thus we can talk of bubbles of air or gas in a fluid *or* of bubbles of fluid containing air or gas. The bubble as such is neither. It is an air bubble in a fluid only from a spatial perspective. Counter-spatially, which means also from the centripetal perspective of the fluid around it, it is a fluid bubble containing air or gas. But what then, of bubbles that are filled with air *and* also surrounded by air – like a balloon or bubbles that children created by blowing into a ring of fluid. These can indeed be said to be rubber or fluid balloons or bubbles, but they still constitute surface boundaries between *enveloped and enveloping* fields and spaces. Any such three-dimensional surface boundary has its own spatial and counter-spatial dimensions. For the surface of a bubble or balloon can either be viewed spatially, as the outer surface of the medium or material it envelops – and counter-spatially, as the inner surface of its surrounding medium, even if this is the same medium such as air. The picture is further complicated by the simple recognition that any boundary, whether of two or three dimensions, has itself two sides. Thus a balloon has both an inner surface that constitutes the immediate boundary of its interior or enveloped field or medium, and an outer surface that constitutes its boundary with an exterior or enveloping field.

Any three-dimensional surface therefore, is not simply one surface but four surfaces in one:

- the inner surface of an outer field
- the outer surface of an inner field
- the inner surface of an inner field
- the outer surface of an outer field

From a spatial point of view there are only two surfaces to an air-filled balloon, an outer and an inner surface. Counter-spatially however, the outer surface of the balloon is the inner surface of an enveloping field – the air around it that is in immediate contact with this outer surface. But the inner surface of the balloon can also be considered as an outer surface of this same enveloping field, being at the inwardly furthest or 'outermost' boundary of the air around it. The result is that, viewed from both spatial and counter-spatial perspectives, we end up with four surfaces, or rather five – for these four surfaces are distinct but inseparable aspects of a singular surface boundary, the balloon as such.

The field-relation between spatial and counter-spatial dimensions of three-dimensional surfaces is basic, not just to an understanding, not just largely of hollow bodies like bubbles or balloons but also so-called solid bodies. Here it is important to recall Plato's insight that form as such are essentially mass-less and immaterial - ideal. You can measure weigh a circular metal disc or sphere, but you cannot measure or weigh circularity or sphericity *as such*. This is why, like the two-dimensional disc in the diagram, which is neither black nor white – and at the same time both – all three-dimensional bodies are reducible neither to the 'matter' that appears to 'possess' their shape or form, nor to the spatial or energetic fields surrounding those bodies – which is just as much constitutive of that shape. Form as such is ultimately something ideal and not material. But even if we consider a pebble on a beach as something material, we forget that is smooth and rounded shape was first rounded and smoothed by a surrounding material medium of water.

The next step towards a Fundamental Topology is to consider both enveloping and enveloped fields not simply as pre-given spatial fields, whether filled or not filled with a 'material' or 'energetic' medium, but as fields of spatial *awareness*. Here the human body – or any body we are aware of in space – provides us with a basic model. A body as we perceive it from without is that body as perceived within a field of extensional, spatial *awareness*. As such, its outer surface is the inner surface of this outer field of awareness – a field that is in no way bounded by our own skins. Similarly both the walls of a room we are in are and the surfaces of any objects or material bodies in that room are also inner surfaces of the same singular field of awareness – on that reaches to the walls of the room and that surrounds both our own bodies and all bodies within that room – or within space as such. As for our own inner field of awareness, i.e. the felt space and quality field of our inner bodily awareness – how we *feel* within our chest or belly for example, and not simply how we feel any organs within it – this is not itself visible from the outside, even if they were to open up our body and look inside it. Only from an external or outer point of view, does the interiority of the human body contain only organs, muscle, fat and bone tissue. We know also that the human body is composed largely of water and that even the molecules of water or any form of matter consist largely of empty space. This external viewpoint therefore makes us little more than skin–bags of water and space.

The outer surface of our skin is the outer surface of this water bag, but it is not the outer surface of our field of inner feeling awareness. The space of this awareness can contract beneath the surface of our skins, expand beyond it, or expand *within* it. The inner field of our bodily feeling awareness of ourselves, unlike the field of our outer spatial awareness, has a counter–spatial and non–physical character – it is an awareness that opens up within an enveloping field and within a psychic skin or 'envelope' of *awareness.*

Counter–space is characterized by its centripetal character, in contrast to the centrifugal character of ordinary space. Bodily skin sensations on the other hand – of warmth or contact, irritation or itching etc. are a counter–spatial awareness of things and people in our outer field – our clothes, the air around us, sound, light and other energies.

Yet our fleshly skin itself is quite distinct from the human organism as a body of awareness with its own skin – its envelope of awareness or 'psychic envelope'. This psychic envelope's surface is not the boundary between the inside and outside of a human water-bag of organs, but between two fields of awareness – the field of our inner bodily self-awareness on the one hand, and the outer field of our sensory awareness of the space and world around us. The inner field enveloped by this psychic envelope is not, however, bounded by it, for – counter-spatially –

it leads *inwardly and centripetally* into that non-extensional or intensional space or field of awareness that is the central theme of this book.

No Fundamental Topology is possible without a fundamental distinction, not only between space and counter-space, but between extensional and intensional space. Counter-space leads into intensional space. But intensional space, far from simply being enveloped by extensional figures and forms surrounded by their own enveloping extensional fields of awareness, actually envelopes these enveloping fields. The extensional spaces of our dreams occupy no extensional physical space but open up and expand within a non-extensional or intensional field of awareness, and are in this sense enveloped by them. What is true of our dreams applies to extensional spaces and fields of awareness in general, all of which open up and expand, within intensional fields and intensional space and are in this sense enveloped by them.

The identification of reality with extension is rooted in the earliest eras of Western thought. To begin with extension was equated with discrete extensional bodies or with a continuous extensional medium – whether fluid or airy or ethereal. It was the Greek 'atomists' who first replaced the ontological duality of Being and Non-being with a cosmological duality of extensional bodies on the one hand and empty space on the other. But Aristotle denied even the possibility of a spatial vacuum free of extensional bodies or an extensional medium, as did Descartes centuries later. And Newton himself could never fully accept the idea that gravity represented action at a distance, unmediated by some form of subtle medium, material or immaterial. Like the Stoic philosophers, Newton's first picture of the cosmos was a stellar realm surrounded by an extra-cosmic void – a realm of pure chaos. To accept an intra-cosmic void or vacuum state seemed, as it did to countless previous generations intrinsically unnatural, in a way summed up in the maxim that 'nature abhors a vacuum'. It also seemed irreligious, for through the identification of created reality with extension, the concept of a vacuum implied also an absence of God within the cosmos. Religious accounts of creation however, themselves posited 'ex nihilo', a nothingness from which the extensional world was created. Fundamental theological as well as philosophical and scientific issues all centred, therefore, on what sort of reality could be conceived beyond a created world or even an eternal and uncreated cosmos if the latter was essentially an extensional continuum. The only answer that seemed conceivable was some sort of extra-cosmic void or intra-cosmic vacuum, a nothingness out of which God created the world or, as in Eastern philosophies, a void which itself gave birth to the world.

Today we have no difficulty accepting the idea of an intra-cosmic void – a spatial vacuum possessing extension but not filled with a continuous extensional medium or ether. General Relativity seemed to make the idea of such an ether unnecessary to account for gravity, replacing it with the notion of curvatures of space itself. But that still left the question of what lies beyond extensional reality, now conceived as a shape of space in time. The lay inquirer still wonders, and rightly so, what sort of reality could have preceded the Big Bang, if the latter represented the birth of space-time. Or, if the entire extensional universe of space-time has itself an extensional shape – a torus or doughnut shape for example – what lies beyond it? The Big Bang gives us a how – provides an account of the cosmic evolution – that seems to fit with the facts but does not explain why the universe should have developed according to certain physical laws and not others. Indeed, Einstein's theory itself allows for universes obeying other laws and taking different forms according to the field-equations relating the geometry of space on the one hand with distributions of matter and energy on the other. Theoretically, a vacuum universe with no matter or energy is defined as flat rather than curved. But other solutions to the field equations allow for curved universes or 'spacetimes' even without the presence of matter or energy as we know it.

At the same time light and gravity have become more of a mystery than ever, the one defining the curvature of space, the other existing as waves filling the vacuum of space and not ultimately dependent on the existence of sources of gravity in the form of extensional bodies. The puzzle of 'dark matter' – the fact that 90% of the gravitational mass of the universe remains unaccounted for – indicates that physics still has a long way to go in attempting to fill the voids in current cosmologies, let alone explaining what they essentially are, and from whence they derive.

Historically, the universe has been conceived of as finite or infinite in extent, continuous or broken up into atoms and collections of atoms separated by empty space, surrounded by an extra-cosmic void or filled with intra-cosmic vacua, flat or curved, spherical or toroidal. But the recurring problem, and the one that cannot be gotten round, is that the origins of an extensional spacetime universe (or of multiple spacetime universes) cannot, in principle, be conceived or mathematically accounted for in purely extensional terms.

What appear as abstruse or intractable problems of physics and philosophy, mathematics and theology are answered by everyday human experience. Aristotle defined a vacuum as a space in which the presence of a body is possible but not actual. And we all know of a non-extensional space which fits this definition – the space of our imagination. We all know too, of an extensional space with no actual extension in space – the spaces of our dreaming. To which we may add the space of our seeing and hearing, which forms no part of any space we see or any objects we hear in it. Perhaps it is not surprising therefore that it was not a physicist but a psychoanalyst - Donald Winnicott - who first ventured a description of non-intensional space not as an empty vacuum but as a 'potential space' in the Aristotelian sense, the space of our openness to possibilities that expands in play and creativity. Long before Aristotle however, it was the Greek thinker Heraclitus who first pointed to the non-intensional space of the psyche itself, and characterized the essential nature of intensional space as unbounded interiority. The term 'psychology' derives from the Greek words *psyche* and *logos*, and Heraclitus's saying, which unites these two words can be considered as the founding statement of any fundamental 'psycho-logy: "You shall not know the limits of the *psyche*, no matter how far you go about it, so deep is its *logos*". The term 'in-tension' means to tend inwards. The inwardness of the *psyche* is an inwardness of its word or *logos*. This is not an ordinary spatial insideness but an unbounded interior space of meaning or inner resonance. If a text is the flat two-dimensional surface of an intensional meaning space, then no degree of curvature of the sheet of paper on which it is printed, no alterations of its extensional geometries - will reveal anything of the fundamental, intensional space of meaning that constitutes the true inwardness of the text. It must be emphasized again, however, that this meaning 'space' is not space in a metaphorical sense. It is fundamental space, understood not as actual sensory space but as the space of potential meaning or sense.

Fundamental Science and Fundamental Ontology

The term 'ontology' comes from the Greek *ontos* – being. Ontology is the Fundamental Science of beings and of Being as such. Awareness can no more be considered a product of phenomena we are aware of than can Being or is-ness be considered a product of beings – of anything or even all that 'is'. Any God or primordial Ground of all beings cannot be considered as one 'supreme' being among others, but only as a primordial field or ground state of awareness. This ground state of awareness does not consist of awareness of any being but is an awareness of Being as such. To paraphrase Sartre, the Being of Awareness is the Awareness of Being. Being as such however, is 'no thing' in particular, no being that is. In this sense it is also Non-being. But Non-being itself is not an empty void. Rather it consists of infinite *potentialities* of being, infinite potential forms and figurations of awareness itself.

The Being of Awareness is an Awareness of these infinite potentialities latent within Non-being. It is an awareness that knows these potentialities in a way ontologically prior to their actual manifestation. The primordial ground state of awareness can therefore indeed be comprehended as a divine knowing awareness that does not follow or reflect a pre-existent universe of beings but is its very source or ground – the source of all beings. Each being within the universe, as an expression of the primordial ground state of awareness, is the actualisation of a potential form or figuration of awareness. As such it also configures its own universal field of awareness. The primordial ground state of awareness therefore necessarily manifests itself not as a single field of awareness but as a field of fields. It is within their own patterned fields of awareness, reality fields in-formed by their own unique field-pattern of awareness, that beings come to perceive each other as phenomena.

The physical reality field is a reality framework or shared reality field, characterized by its spatio-temporal or extensional character. Within it, beings appear as bodies in space-time. Both physics and physiology, physicist and physician, however, share a common conceptual framework derived from the physical reality field. This conceptual framework is an ontological framework of a particular sort, one which reduces beings as such to bodies in space-time, and understands awareness itself (Greek *noein*) as a product of the interaction of bodies in space-time, and regards knowledge as something deriving not from familiarity with beings (*gnosis*) but from observation of bodies.

From the point of view of the physiologist and physician it is bodies and brains that see and hear, think and feel, breath and metabolise. From the point of view of Fundamental Science it is quite the reverse. It is not bodies and brains but *beings* that see and hear, think and feel, breath and metabolise. We do not see because we have eyes, hear because we have ears or think because we have brains. We have eyes, ears and brains because we are seeing, hearing and thinking beings. Similarly, we do not breathe because we have lungs. We have lungs because we are breathing beings. What we inhale as beings is not physical molecules of air but the living medium of our very existence as beings. This is the 'life-breath' of awareness that in Greek went by the name of *psyche*, constituted not from quanta of energy but from qualia – qualitative tones and intensities of awareness.

From the point of view of the physicist, bodies as such are extensionally bounded units existing in a pre-given space-time field. From the point of view of any being however, not least the human being, bodyhood is nothing essentially bounded nor does it consist of a measurable volume in space. Every body, appearing within my own extensional field of awareness, is my body. For as Heidegger pointed out, when we point at a windowsill our bodyhood is not a volume enclosed in skin, nor does it merely reach to the tips of our fingers. It extends to the windowsill itself, as a phenomenon within our own bodily field of awareness, and one that we are subtly aware of in a bodily way – not just visually, but in a tactile and even aural way. For in our 'seeing' of the windowsill we also know what it would feel like to touch in comparison with the pane of glass, what sort of sound it would make if struck. Indeed on a certain level of awareness we already hear the texture of its wood as a particular tone or timbre.

What if not just human bodies, but material bodies of all forms, from atoms and molecules, to planes of glass and windowsills, are in essence embodiments of modes of awareness – not human perceptual or conceptual awareness but a type of natural pre-conceptual, pre-perceptual and indeed pre-physical awareness that is the foundation of all forms of consciousness. As such they must be considered as aware beings and not simply as bodies which we human beings are aware of. And as such, they will also configure their own patterned fields of awareness, though not necessarily of an extensional, spatio-temporal character.

Our distance or closeness to one another as human beings is not something measurable by the distance that separates us as bodies in space. We can be close to another human being though there are thousands of miles separating them from us. Similarly, we can be physically close to them whilst at the same time being 'miles away'. The distance or closeness of beings – and not just human beings – is an intensional distance that is not measurable in extensional terms. But it is no less also a felt bodily closeness or distance for that reason. When we feel close to a loved one who is far away we feel that closeness in a tangible bodily way – as a warmth of feeling within us for example. And yet this warmth is not itself a measurable physical warmth – our temperature does not increase as it might do hugging that person physically.

Fundamental Science demands and allows us to think the hitherto unthinkable. That as beings, material bodies are no more extensionally bounded than our own bodyhood is in relation to a windowsill, a pane of glass or another human being. That in this sense, material bodies are not separated from one another in extensional space, nor do they move in space. That movement as such is not simply a change of place but *kinesis* in the sense that Aristotle understood it – a change of the state of any form whatsoever. And that space as such is not a uniform system of coordinates in which any body can occupy any position. Rather, every body, as the Greeks already understood it has its own natural place or topos, as did temples and the people that visited them, as did gods and mortals, as do minerals, plants or animals, planets and stars. These fundamental considerations may be thought of as too deep and philosophical to have any direct implications for our current physical–scientific understanding of the universe. And yet they do, for two reasons. Firstly, they provide a new and meaningful account of the whys and wherefores of new scientific concepts such as non–locality, infinitely extended matter waves etc., which challenge the traditional understanding of material bodies as bounded units. Secondly, these fundamental considerations open up new ways of understanding such basic concepts as space and time, distance and movement, mass and energy, light and gravity.

If, using an extensional analogy, every body has its own place, and like a temple standing out in a Greek landscape and lending a particular cast to that landscape, configures its own 'space' or surrounding field – then any idea of a movement of bodies in space must give way to an understanding that the movement of bodies is in fact a movement of spaces.

If all movement is essentially *kinesis* – not change of place but change of state or form, then 'energy' is not simply some 'thing' that conserves itself in every transformation. Instead, it is the formative and transformative activity (*energein*) through which an inexhaustible field of different potential field-patterns takes shape in space-time, appearing as energy potentials and kinetic energies.

And if the mass of an extensional body is the expression of a high field-density of potentialities, manifest as potential energy, then maybe gravity needs to be considered in a different way – not as a force exerted by a mass, nor as a function of matter in motion, but as that which first draws or gathers formative energy potentials into manifestation as densities within patterned spatio-temporal fields.

Then we can cease to see material bodies as pre-given entities but rather as the spatio-temporal extensionalisation or 'bodying' of densifying intensional fields which lay out extensional spaces within them as light and gather them gravitationally as bodies.

Gathering' and 'laying out' are the root meaning of the Greek verb legein – that which first makes possible any ordered numerical counting and verbal recounting of things, or any postulation of an inner order or logos within the outer universe or cosmos. Hence the saying of Heraclitus: "Listen not to me but to the logos".

Fundamental Science and Causality

The basic principle of field-dynamic phenomenology, namely that no event can be explained by other events occurring in the same field of emergence, appears to contradict the basic principle of causality, and in this way to offend both common sense and scientific convention. Rain, surely is a cause of people getting wet, yet we do not tend to say 'the rain caused me to get wet'. Rather we say 'I got wet in the rain'. The event we speak of is not rain falling or getting wet, but rather 'getting wet in the rain'. It is true that in ordinary discourse causality may be spoken of directly, as when we hear on the news an explosion was caused by a bomb. But what does this explain? For tracing back a linear chain of causality – from explosion to bomb to perpetrators etc. in no way necessarily helps us to understand the cause of the explosion in a more fundamental way – why a bomb was placed in an aircraft, why it was not discovered. Causality gives us an explanatory 'what' but not a fundamental reason or 'why'. Nor does it even give us a 'how'. Biologists can describe in detail the chain of events through which a virus infects the body and triggers an immune response which is experienced through disease symptoms. But in what sense does this allow us to regard the virus, or bacterium, or cancer cell as the fundamental 'cause' of the illness? Why do certain people not contract diseases during an epidemic? Why, if potential pathogens are present in our body all the time, do we only get ill from them, if at all, at certain times and not others? We can invent other explanations for such phenomena, which in turn are adduced as independent causes – stress or other factors, for example which are said to 'cause' a weakening of the immune system. But then why, for example, are certain people vulnerable to stress? Again, more 'causes' can be adduced, each of which in turn raises new questions requiring more sophisticated causal explanations. The paradigm of causal explanation is never itself questioned, for on it depends the much acclaimed power of science to predict events and attribute causes to them on the basis of statistical correlation alone – even though it is admitted that such correlations are not in themselves evidence of any sort of causality at all. Medical science rests on such correlations however. And the physician has no hesitation in claiming that they provide evidence of causality – claiming that smoking, for example, has been proved to 'cause' cancer or heart disease.

From a field–phenomenological perspective, causal explanation fails to recognize the essential nature of the events and phenomena that we seek to explain. As Heidegger put it "All explanation reaches only so far as the explication of that which is to be explained." Thus the genetic explanation of illness already implies a certain concept of what illness as such essentially is. "To be in a position to explain an illness genetically, we need first of all to explain what the illness in itself is. It can be that a true understanding of the essence of an illness...prohibits all causal-genetic explanation....Those who wish to stick rigidly to genetic explanation, without first of all clarifying the essence of that which they wish to explain, can be compared to people who wish to reach a goal, without first of all bringing this goal in view."

Going back to our everyday example, we can explain a phenomenon such as 'wetness' by attributing a cause to it such as 'rain'. Or we can define the very thing to be explained in a way that includes its broader field or context of emergence. In this case for example the phenomenon to be explained would not be wetness 'caused' by rain, but 'getting wet in the rain' or more broadly still 'getting wet in the rain while running to the station after waking up late with a hangover from the night before'. Notice that the linguistic paradigm here is not the Noun-Verb-

Noun paradigm represented in propositions of the form 'A causes B' or 'one thing causes another thing' (for example 'rain causes wetness' or 'smoking causes cancer'). Rather the event is described dynamically, through the use of the present participle of the verb, the '-ing' form. The event itself is not understood in the manner of a causal proposition with a subject and object noun. As a result, its 'why' is not a cause, not even the possible 'cause' of 'getting drunk in the pub the night before'. For all such events are also events occurring in an unstated context that includes not only previous events but parallel and probable events, synchronous events and alternate possible events and even patterns. The 'why' or 'because' of the event of getting wet or getting drunk is not a causal 'why' or 'because' in the ordinary sense. It is a hermeneutic 'why' or 'because' – carrying a meaning that derives from a broader context of emergence. Fundamental causality takes the form of causal patterns of events in both space and time in which no one event can be singled out as the cause of any other. Instead all the events in the pattern derive their meaning from the overall pattern of events that constitutes their context of emergence.

Field-patterns of events carry their own significance – they are field-patterns of significance that transcend narrow causal explanation. Understanding this is of no small significance in itself. For causal explanation is central to medical science, which, to its scientific discredit, acknowledges no meaning or significance in illness beyond its supposed 'cause'. Little or no attempt is made to understand the patient's symptoms as a text emerging within the larger context of their lives and relationships. Instead they are read only as diagnostic 'signs' of some possible cause.

So crime, drug use or terrorism, are seen as social illnesses to be cured by eliminating them or identifying 'causes' for them –

rather than understanding the social and economic *contexts* in which they manifest – a context being something quite distinct from a causal agent. Contexts are constituted by patterns of interrelationship. So along with the displacement of contexts by causes goes an artificial separation between the health of the individual or of society as a whole and the health of human relations *between* individuals and within social groups, communities or corporate bodies and institutions.

The question of how patterns of events emerge is entirely sidetracked by the search for 'causes'. They emerge in the same way as patterns of words in a sentence or text, or patterns of events in a dream – as expressions of inner field-patterns of significance. That a particular word has a high frequency of occurrence in a particular text is no evidence of that word 'causing' that text, or 'resulting' in certain features of it. Similarly, the fact that a repeated nightmare contains the same dream figure is no evidence of that figure 'causing' the nightmare.

Nowadays, of course science speaks of multiple causal 'factors' and not only of single causes. In place of 'multi-factorial' analysis of causality, field-dynamic phenomenology introduces the concept of a multi-field and a field-interactive understanding of the process of emergence of events. Any given phenomenon or pattern of events emerging in a given field is understood as the expression of multiple, overlapping and interacting source fields. A Fundamental Epidemiology for example, would recognize that the emergence and spread of infectious diseases has to do not only with the transmission of pathogens from one individual to another but with interactions occurring between the inner field-pattern or morphic field of the pathogen itself and that of the organism. It would also recognize that the susceptibility of an organism has to do not only with its biology but with its place in a variety of relational fields – environmental, electromagnetic, social. In the case of human beings, for example, the cultural and social fields of the individual are paramount. The dominant diseases of a particular culture, whether tuberculosis, hysteria and neurosis, or AIDS and cancer, have always had, as Susan Sonntag has shown, a highly specific symbolic significance within that culture. This she sees as a danger to medical objectivity, objecting strongly to replacing a causalistic medical-scientific approach to the diagnosis and treatment of disease with one based on an understanding of 'Illness as Metaphor' (the title of her book). Her arguments however, fail to acknowledge the degree to which the causal explanations invoked by medical science itself are not based on biological facts so much as on ideological metaphor. These metaphors are then taken as literal facts. Genetics, for example, is based on the dominant metaphors of information technology and computer languages, treating DNA as an instructional code or language with its own letters and words.

Immunology is based on military metaphors of a body 'fighting' and 'annihilating' 'foreign bodies'. Just as Hitler saw the Jews as a social bacillus or cancer affecting the body of the Volk, so does immunology see all illness as the work of malignant 'foreign bodies' or 'antigens'.

The entire 'science' of immunology, for example, is founded on the metaphor of 'Illness as War' – a battle between 'self' and 'non-self' cells and molecules, one in which an 'arsenal' of immunological 'defences' is employed and 'armies' of immune 'defenders' are mobilized to eliminate invading antigens:

"[W]hen immune defenders encounter cells or organisms carrying molecules that say 'foreign', the immune troops move quickly to eliminate the intruders."

"The immune system stockpiles a tremendous arsenal of cells. Some staff the general defences, while others are trained on highly specific targets." (from a teaching material on immunology).

Sonntag also ignores the effects of the patient's own beliefs about illness on their interpretation of symptoms, and the role of medical beliefs in encouraging patients to 'somatise' – to embody a felt sense of dis-ease in a way that manifests as a recognizable disease symptom. As for the assumed effectiveness of causalistic 'scientifically' based medical treatments in combating and defeating disease, this is far from being an unquestionable truth. For the facts are that:

- Medical treatment itself is the single largest cause of death after coronary disease, stroke, cancer and AIDS.
- The life-expectancy of untreated cancer patients is higher than that of treated ones.
- There is no statistical evidence for the effectiveness of costly intensive care units.

"Neither the proportion of doctors in a population nor the clinical tools at their disposal nor the number of hospital beds is a causal factor in the striking changes in overall patterns of disease." (Illich)

As Illich has pointed out, most major infectious diseases such as tuberculosis declined in the 19th century – due to improvements in wages and nutrition – and well before the germ theory of disease causation was invented and antibiotics discovered. Most cardiovascular diseases and cancers are not caused either by viruses or defective genes. Only extremely rare diseases are clearly linked to defective genes. Yet, even here, no explanation exists for how these genes 'cause' the diseases. For the relation between genes and protein production is a reciprocal or dialectical relation and not a relation of one-way causation.

Fundamental Science and 'Free Energy'

The subservience of science to technology is the expression of the role of this technology in capitalist economies. It was Marx who first showed how changes in technology – in mankind's relation to nature – brought with them changes in social relations. Feudal society was brought down not by revolutions but by the development of technologies that laid the foundation of manufacturing industry, and a new economy incompatible with the old pattern of social relations. Today, alternative scientific models marginalized by the scientific establishment pose no less a threat to capitalist economies. That is because they bear within them the seed of new technologies that challenge the whole capitalist economic model of supply and demand. The latter is predicated on the idea of inherently limited resources, above all energy – the foundation of industry.

Today, the alternative scientific models applied in so-called New Energy Research offers the very real prospect of mankind one day being able to draw upon unlimited free energy from a Fundamental Field. This primordial 'vacuum' field, far from being an empty or entropic void, is, as physics itself is beginning to recognize, a fullness or pleroma. That it does not manifest as such is due to its essential nature as a field of potential or 'zero-point' energy. John Davidson nicely compares this energy to that of a rope in a balanced tug-of-war competition, explaining that the forces exerted by both teams are by no means zero even though they may indeed sum to zero and not be apparent. He compares the primordial vacuum sheet itself to a tightly stretched piece of paper. What appears to us as energy and matter are vibration waves or ripples on this sheet which can also take the form of apparent particle spin. His reference to tensed or stretched ropes and sheets has a deeper significance in relation to the basic terms of Fundamental Science. The words 'tension', 'extension' and 'intension' all have their root in the Latin tendere – to 'stretch' or 'span' – as do the verbs 'tend' and 'attend', 'extend' and 'intend', tender and tense. The word 'tone' has a similar root in the Greek teinein – also meaning to stretch. It is from a stretched string that musical tones are produced in different intensities. Strange therefore, that the concept of toned intensities as such is not one of the fundamental concepts of modern physics. In contrast, Fundamental Science understands the fundamental reality as something composed of patterned tonal intensities, the latter being the music of the inner universe.

The distance between a source of sound and a listener can be measured, as can a wavelength of tonal vibration, or the distance between two nodes of a stretched string. But the experienced distance between two tones as such is no more measurable than the distance between two colours, the distance between two moods or tones of feeling, or the emotional distance between two human beings. That is why in music it is described as an 'interval' rather than a distance. Fundamental Science understands the space of our resonance with a piece of music as something fundamentally distinct from the extensional physical space in which tones manifest as vibrations of air particles. Furthermore, it understands extensional space itself as an expression of a more primordial type of space – an intensional space in which distances are comparable to tonal distances or musical intervals. The fundamental essence of space has a temporal character.

Fundamental Space is a patterned field of co-presence composed of different tones separated by intervals and amplitudes or degrees of intensity. Fundamental Time, on the other hand, cannot be tautologically defined as a succession of tonal intensities or patterns in time. Instead time itself is a function of successive tonal intensities and patterns. The deepest expression of this truth is found in music. A melody is not a succession 'in' time of notes or tones of longer or shorter duration, and of lesser or greater volume or intensity. The melody itself is a tonal shaping of time. Fundamental Time is a succession of different tonal qualities and intensities of awareness, whether experienced through music or through the successions of moods or field-states of awareness that constitute our everyday life. A tone does not persist 'in' time. Time itself is a function of patterned tonal durations and intensities of awareness – weakening patterns and intensities being experienced as 'past' events receding like the sounds of a passing car, and increasing intensities being experienced as events nearing our present like the sound of an approaching car.

The fundamental concept of intensities is given recognition in New Energy Research through the importance it attaches to 'scalar' values, defined only by magnitude or intensity and not by directions or vectors. From the point of view of scalar electromagnetic theory, two or more electromagnetic fields may sum to zero but conceal zero-point energy – potential energy in the form of scalar potentials hidden within apparently entropic ground-states in which no measurable field effects or field-patterns manifest. We might compare these hidden energy potentials in the vacuum field of 'empty' space with silences in music; silences that are not empty or vacuous but pregnant, audible not as physical sounds but as inner sounds – as resonant tones of silence. As any appreciator of the symphonies of Anton Bruckner knows, these tones of silence can be rich and more loaded in power and intensity than the grandest of climactic chords that emerge from them.

The terms 'power', 'potency' and 'potentiality' all share a common root. We think of energy as a source of power, something which has the potential to do 'work' – to push bodies or charges around. In doing so however, we reinforce another Fundamental Misconception. For power, properly understood, is potentiality. Conversely, potentiality is power – whether this power be experienced as potentialities latent within oneself, made visible in the vastness of the cosmos, whether it be understood as the creative potentialities of awareness as such or the creative power of a hidden God. And energy, properly understood, is not a source of power but a fundamental expression of potentiality, the formative and transformative activity that translates potential into actual intensities. Thus the very term 'potential energy' is something of an oxymoron. Here again, it is through questioning scientific language itself that we come to the hub of the matter – in this case the relation between Fundamental Science and the 'New Science' of 'free energy'.

Fundamental Science and Entropy

One of the most fundamental misconceptions of conventional physics is the identification of entropy with disorder or lack of form. High entropy, in physical-scientific terms is understood as a state of maximum energetic equilibrium, associated with a state of minimum thermal and kinetic energy (the second law of thermodynamics). But as Rupert Sheldrake has argued so coherently, if entropy implies disorder, why is it that a salt solution crystallizes when placed in a cool environment? The evening out of the temperature of the system as the solution cools and loses its temperature differential with its environment constitutes an increase in entropy but goes together with the increase in order or formal complexity constituted by crystallization. Similarly, as high energy plasma cools, the electrons and atomic nuclei that compose it come together to form first gaseous atoms, then molecules, then fluids and finally solid crystalline structures and 'supermolecules'.

In physics, stability is associated with states of minimum potential energy – a ball poised on the top of a mound for example, having higher potential energy than one that has rolled to the ground but lesser stability. The electron shells surrounding atoms have different quantum energy levels, the outermost ones with the highest levels exhibiting the least stability. But quantum physics has no way of predicting the form of complex molecular structures on this basis. That is because, as structural complexity increases so do the number of different possible structural forms sharing the same level of energy and stability.

Following Waddington, Sheldrake pictures these different possible forms as landscapes – rather like three dimensional mountain ranges down which water flows, or mounds of different heights and shapes down which a ball can roll into canals between them. The important thing is that the mountains or mounds can descend into valleys or canals at the same height, allowing water to flow into more than one river or stream, or a ball to roll along more than one canal with the same energy. Any canal between two mounds can branch out into others at the same height, into either of which the ball can roll and come to rest.

It is a well-known fact that when a magnetized bar of iron is heated above a certain temperature, it loses its magnetic properties and polarity due to the randomisation of atomic movements and spins. What is less well known is that when it cools below this temperature, it polarizes magnetically but does so in an unpredictable way – each end of the bar becoming either a magnetic north pole or a south pole. This is an example of two potential field-patterns of emergence existing below a certain ground state represented by a particular scalar value (in this case the 'Curie temperature') and comparable to two canals or valleys in a 'morphogenetic' landscape.

We have here a clear example of the fundamental nature of entropy, a word which derives from the Greek 'trope', meaning a change or turning point. The term 'entropy' was coined to signify a state in which there is little potential for change. The higher the entropy, the less potential energy and the less potential for change. Understood in a more fundamental sense however, entropy is en-tropy or extra 'tropy'. It is an increase in formative potential and thus an increase in 'potential energy' in the fundamental sense of *energein* – formative activity. Because of the conventional identification of entropy with lack of form and low potential energy I coin the term 'intropy' to describe this complementary dimension of entropy as 'extra tropy' – a turning point with many possible outcomes.

The Eleventh Fundamental Distinction

The eleventh fundamental distinction is between entropy and intropy. Entropy is a decrease of potential energy. Intropy is the increase of potential energy as 'inergy' or formative potential. Entropy reduces the degree of patterned differentiation of intensities within an energetic field. Intropy increases the intensity of different potential field-patterns that can emerge into actualization. A ground state of minimum potential energy and maximum entropy is at the same time a state of maximum intropy – a maximum intensity of inergetic or formative potential.

Fundamental Sounds

External, physical sounds are patterned shapings or envelopes of longitudinal wave vibrations. Inner or intensional sounds are patterned shapings or envelopes of inner tonal intensities of awareness. Every inner field-pattern of awareness is an inner sound, being a shaped configuration or pattern of inner tonal intensities of awareness. Inner field-patterns of awareness are both inner field-patterns of significance and inner sounds – patterned tonal intensities of awareness.

In ordinary language, the intrinsic significance of word-sounds passes us by. According to conventional linguistics, word-sounds possess no intrinsic meaning or sense at all, but are arbitrary signifiers. Were they not, so the argument goes, all languages would use the same sounds to signify the same meanings and thus share similar or identical vocabularies. This argument rests on a basic misconception – the idea that linguistic meaning is a function of verbal signifiers rather than an expression of felt sense and inner field patterns of significance. It also reflects a basic confusion between sense and signification. Inner sounds do not indeed possess, signify, 'point' or 'refer' to determinate meanings. They do not signify meanings in the way that words do, so much as they directly sonate or sound out felt senses that are wordlessly experienced. In doing so they give physical form to inner field-patterns of significance, allowing them to re-sound or re-sonate as physical sounds or the sound patterns of words themselves.

The audible sounds we shape and utter with our bodies when we speak are the echo of inner sounds with which we utter our bodies themselves. A simple consonant sound such as an extended 'Mmm" can be thought of as communicating a variety of senses such as 'I understand' or 'This is delicious'. In essence however, it has no verbally definable meaning at all, for rather than signifying any such meaning it directly bodies forth a felt sense that is indefinable in words, irreducible to any verbal signifier or signification. This same is true of interjective vowel sounds such as 'Ah' or 'Oh', or 'Eh? Within any given language, particular interjective sounds do indeed attract and acquire conventional significations, but the latter draw from a field of potential senses that transcends these actual or probable significations. That is why the meaning of words as sound patterns can never be reduced to their meaning as words. The resonance or 'felt sense' of particular word-sounds does not have its source in verbal significations – in the given meaning of words containing that sound. Rather the opposite: word sounds carry felt senses that link and at the same time transcend the given meanings of the words containing them – in whatever language. It is only because these felt senses cannot themselves be pinned down or defined in words that sounds are seen as lacking any intrinsic sense.

The fundamental sense of a given word sound is an inner resonance that is felt in a wordless, bodily way – necessarily so, since the uttered sound is itself the bodying of an inner sound, the shaping of a felt tonality of awareness or 'feeling tone'. Just as speech rides on patterns of vocal intonation, giving phonetic shape to specific tones as word-sounds or phonemes, so does verbal communication as such ride on patterns of inner tonation. These are wavelengths of inner tonality or feeling tone, whose shaping or modulation as inner sounds enables them to act as direct 'telepathic' carrier waves of felt senses. Telepathy is not essentially the transmission from one mind to another of a signifier – a word or mental image, but the transmission of felt senses and felt patterns of significance through inner sound and feeling tone. We do not tune in to others through understanding their words. We understand their words in so far as we are already 'tuned in' and on their 'wavelength'. These are not mere metaphors. Rather, outward, verbal communication itself is the metaphorical expression of an inner communication that rides on feeling tone – on wavelengths of resonant attunement.

The human organism is the instrument or organon which gives bodily form to inner feeling tones by translating them first of all into muscle tone. That is how it is also able to translate felt meanings or intents – intentionality — into those patterns of muscular in-tensionality which prepare the body to move or speak. Just as speech involves muscular activity and movement so is movement itself a form of speech. But non-verbal communication and 'body language' no less than verbal communication and language, are both expressions of a deeper level of direct intensional communication whose medium is intentionality as such – the felt intent of a speaker. Intent is both an attunement to inner resonances, tonalities and patterns of significance, and thus central to listening – and the inner modulation of those tonalities on which patterns of speech and verbal communication ride.

Field Envelopes of Awareness

The infant does not hear the sounds of things. It hears things as sounds. It does not hear a given sound as a 'clock' ticking, a 'car' passing by, an approaching 'footstep' on the stairs, or the sound of its mother's 'voice'. Rather the converse – what the car, footstep or mother is, is something experienced by the infant through and as its sound, its tone and timbre of vibration. Nor does the infant see a 'clock' on the wall or a patterned 'wallpaper'. Rather visual objects too are experienced as sensual shapes that are intrinsically meaningful even without having verbal labels and functional significances attached to them. These shapes are comparable to silent sounds. For it is the tonality of an object's colour and the vibrational timbre of its material texture that the infant experiences as intrinsically meaningful – in the same way that an adult might experience a sunset, spring breeze, or the feel of a lover's flesh. Felt sense is meaning sensed in a bodily way. On the one hand it is distinct from felt bodily sensations. On the other it is something inseparable from sensation. For the infant, bodily sensation as such is something intrinsically meaningful, but not in the same way as it is for the adult, except in special circumstances. For this meaning does not take the form of phenomenal signifiers – of formed concepts or percepts, words or even verbal images of things such as cars. Neither are 'things' experienced as 'out there' in the world as opposed to 'in me'. In its ground state of awareness, the infant's field of inner bodily self-awareness is a medium of pleasurable sensual streamings which constitute a medium of inner vibrational contact with the world – a direct relation between its own withinness and that of the things and people around it.

Much emphasis is placed in psychotherapeutic and psychoanalytic literature on the early body contact of mother and child, in particular the importance of skin contact, holding and the infant's relation to the breast. But this is a form of extensional contact through a mutual outer surface. The nature of the infant's earliest or 'primary' contact with the world is not extensional but intensional. It experiences itself within the things it sees and hears just as it experiences them within itself. Intensional contact is not mutual contact through an outer surface but a mutual withinness experienced through the medium of inner vibrational touch. For the adult, this type of intensional contact is experienced primarily through speech and music – to understand the sounds of speech and music is to experience a mutuality of withinness – finding ourselves no less within the sounds as they are within us.

When a baby leaves the mother's womb at birth it continues to dwell in the vibrational medium of its own intensional body – an envelope of awareness filled with the fluid medium of feeling tone, and composed of the rich tones and textures of its bodily awareness. But the boundary of this envelope or body of awareness is by no means identical with its physical, skin boundary. It is experienced, like the waters of the womb as an enveloping 'sonorous field' or 'bath' (Anzieu/Lecourt), a vibrational envelope whose outer skin or circumference extends as far as the infant's own immediate sensory environment itself. But the infant does not use its bodily senses to see and hear, feel and touch things in this environment. It feels itself within the things it senses in its sonorous field and feels touched by them inwardly. The actual sound vibrations or shades of light and colour that emanate from things are only the outer sensual medium for its primary sense – contact through what Seth calls 'inner vibrational touch'.

The root meaning of the words 'woman' and 'wife' is 'to vibrate'. The waters of the womb are a sonorous bath, a fluid vibrational medium in which the foetus hears the mother's heartbeat and voice, as well as sounds coming from her physical environment. Its own body is all ear. After birth the infant still floats in the larger body of its own vibrational envelope and its bodily self-awareness is an awareness centred within this larger body – its own 'womb body' or 'mother body' (Mutterleib or mother-body being the German term for 'womb').Within this body it feels itself weightlessly borne or carried even as it did in the mother's womb.

Within its vibratory envelope of awareness the infant experiences no extensional distance or spatial separation between its own physical body and an external source of sound. Its awareness of its own physical body surface or envelope as a bounded volume in space arises only through physical contact with things and people – in particular of course, the mother. Surface skin sensations constitute a second skin or envelope of awareness, one which serves as an interface between an outer field of sensory awareness, and a contracted inner field of awareness now experienced as something bounded by its own physical body and contained within its own skin. It is in developing this second, physical envelope of awareness – and feeling comfortably contained within it – that physical contact with the mother and others is important.

The infant's sense of its own spatial separation as a body from the other bodies around it, and of space itself as something empty and void rather than fluid and full comes about through the alternation of contact and separation with the mother, and its relation to the inner void experienced both as thirst for milk and as contact hunger for the mother. It must be emphasized however that contact is not 'merger' but the experience of a containing surface, skin or envelope that simultaneously distinguishes and unites self and other. What psychoanalysts consider as an infant's primary and original sense of 'merger' with the mother is not merger at all but intensional contact – the infant's experience of inner contact or relatedness as well as outer contact. Actual physical contact with the mother serves primarily to protect the infant from impingement by outer or inner sources of dis-ease or distress – from overstimulation on the one hand or understimulation and the experience of an inner lack or outer void on the other. Its primary role is to comfort the infant by maintaining what Winnicott called its sense of 'going on being' – protecting the infant from environmental impingements which disturb its ground state of awareness and disrupt its vibratory envelope. It is through the latter that the infant remains in a state of nourishing inner contact with its environment rather than perceiving it as a spatial field separating it from the things and people around it. Where the mother's physical contact with the child is not itself an expression of the inner contact she feels with it, the infant will cease to feel this inner contact too.

The intensity of the infant's 'separation anxiety' and the degree to which it experiences extensional spacetime as absence of emptiness is inversely proportional to the degree of inner contact it feels with its environment and with the mother herself. Hence the importance of the mother's own ability to make inner contact with the infant through her physical presence and contact with it – and to sustain this inner contact inwardly even as she withdraws contact or withdraws from physical presence. The difference between extensional physical contact and inner or intensional contact is the difference between the mother's external relation to her baby as a body and her inner relation to her baby as an aware being. She can see, hear, hold, touch and feed her baby's body without seeing, hearing, touching and feeding it as a being – the dimension of inner or intensional contact. Similarly, for the infant there is a difference between experiencing itself as a body on the one hand and bodying itself on the other – experiencing its physical body as an embodiment of its own being. To be seen and heard, touched and held as a being – and not just as a body – is vital here. So too, however, is the development of a third envelope of awareness – what I call the phonic envelope.

The infant expresses sensations of dissonance, displeasure and distress in sound. In doing so however, it is not bodying these sensations but evacuating them. It does not make sounds with its body but finds itself uttered as sound by its own body. The phonic envelope is the instrument with which the infant learns to utter its own body – gives bodily form to its own inner feeling tones through its face and eyes, sounds and movements. It consists of inner sounds – felt shapes or envelopes of feeling tone that find expression as fundamental sounds – simple phonemes and syllables associated with the babble of babies. The word 'infant' derives from the Latin in-fans or 'non-speaker'. The recent acknowledgement given by 'scientific' studies to the importance of 'baby talk' concentrates only on its role in helping the infant to learn basic word sounds and thus to speak. Such studies focus only on the outward aspects of the parent's baby talk – the vocal elongation of vowels or the emphatic enunciation of consonants. They completely miss the fundamental scientific significance of these word sounds as outward expressions of inner sounds through which the baby learns to feel and shape its own phonic envelope. For it is only through the latter that the infant learns to quite literally 'utter' its own body from within, thus experiencing it as a shaped embodiment and expression of its own inner feeling tones. It is through the phonic envelope that the baby becomes a 'person'.

In baby talk, the mother does not merely mirror her baby's facial expressions or echo its babbles. She allows herself to resonate with the feeling tones expressed in the sounds it makes.

Her own mouth shapes and facial expressions, as well as the sounds she makes, show the baby how to give a distinct phonemic and facial form to these feeling tones. The baby responds principally not to the mother's sounds but to her facial mask or persona – that through which she sounds (per-sonare). A letter is the silent face or look of a sound. The face of the mother is itself the silent look of a sound, the face of an inner sound – communicated as much by the felt tonality of her gaze as by the shape of her mouth and the sound of her voice. The latter do not only communicate her own feeling tones nor do they simply mirror or echo those of the baby. They give form to these feeling tones, and in doing so both amplify them as the very medium of resonant inner contact and communication between mother and child.

The metamorphic mobility of the mother's facial expression, her mouth, brow and eyes, and the tonality of her gaze as well as her voice are central in helping the baby to become a person, i.e. to express and personify its feeling tones in a bodily way. They express her ability to re-sound or resonate with the baby through the medium of feeling tone and at the same time give phonic and facial form (morphe) to this resonation. The morphic resonation of mother and child is not practice for 'proper' verbal communication. It is a fundamental medium of communication in itself. What the baby learns through it is not to speak but to shape its own 'sound body' or phonic envelope, creating shaped envelopes of feeling tone in resonance with their own formed facial and vocal expression. Every basic consonant such as /b/, /m/ or /l/ is a fundamental shaping of the phonic envelope or body of feeling tone that constitutes a vital membrane of the human organism as such – the instrument or organon with which we body feeling tones in muscle tone, speech and movement.

In silently shaping an /m/ sound for example, the infant learns to experience its body as a containing womb, membrane or amnion. Through shaping its mouth in preparation for a stop sound such as /b/ or /p/ sound it allows the membrane to swell in preparation for a vowel sound to burst out. The infant's oral cavity serves, like the womb itself, as a mouth of creation and a microcosm of its organism or phonic envelope as a whole, its inner felt sense of bodily boundedness or openness. When the mouth is opened to release vowel sounds, we sound out our own inwardness, baring our souls in such a way as to give them a new body, not one of flesh and blood but of sound – ensouled shapes of breath and tone. The very relation of outwardness and inwardness, body and soul, form and feeling tone is expressed phonically in speech as a relation of consonant and vowel.

If parents are disinclined to engage in baby talk or if the infant's sounds and syllables are prematurely interpreted by the parents as incipient words then they are not appreciated as shaped expressions or personifications of feeling tones. This being the case, the infant acquires language and a mind – its verbal envelope – in the absence of a phonic envelope. As a result, however, it loses a felt sense of the meaning of words themselves, and a capacity to understand verbal communication as the vocalisation of felt meanings and feeling tones. The verbal envelope serves only as a tool of manipulation, signifying things 'out there' rather than expressing things 'in here'. The child does not learn to dwell in language – to experience it as a second skin or body filled with inner meaning and resonances that are themselves felt in a bodily way. Nor does it feel inner meanings or resonances in the very sounds and rhythm of speech. The so-called mind-body split, just like the split between 'persona' and 'self', 'ego' and 'self', 'false self' and 'true self' – in other words the outer and the inner human being – has its roots in the split between signifiers and felt sense that develops in the process of infancy and language acquisition.

Fundamental Science and Mysticism

In many spiritual traditions, from Kabbalah to Vedic philosophy, great mystical significance is attached to specific letters or sounds or their combination in god-names and mantras and this has a genuine scientific basis in both the fundamental nature of sound and in the nature of fundamental sounds. The words 'mystic', 'mystical' and 'mysticism' all derive from the Greek word for initiates: mustai. The literal meaning of mustai is 'closed-mouthed ones', those who bear the secrets of the universe in silence within them. But the 'mystical' secret of this word lies in its first syllable – mu. For this syllable both names a letter of the Greek alphabet (③) representing the 'm' sound, and constitutes a Greek word in itself. As a word it denotes, paradoxically, a wordless sound, which like the 'm' sound, is made with the mouth closed – a humming sigh or groan.

It is no accident that so many Vedic and Buddhist names and mantras (BRAHMAN, AUM, HUM etc) contain the closed-mouthed 'm' sound named in Greek by the syllable mu, and that, in Kabbalistic mysticism is considered one of the three 'mother letters' of the Hebrew alphabet. If we think only of the lexical field of English words containing this sound we find word groups that seem to indicate a common realm of meaning – for example womb, chamber, embryo, amnion, membrane, umbilical, mother, matrix, emerge, swim etc. Such groups do not necessarily consist of etymologically related words, nor does the existence of such etymological groups explain the role of the particular sounds that are common to them. The mouth itself is a type of womb or resonant chamber from within which we give form and birth to distinct speech-sounds or phonemes.

Specific word sounds such as /m/ possess an inner resonance or felt sense that transcends verbal definition. This resonance or 'inner sound' is echoed in words or word groups containing that speech sound but they are not identical with it. Instead the word sound itself, whether vowel or consonant, gives form to the fundamental inner sound. The latter cannot be audibly uttered, chanted or sung, although prolonged chanting or singing of words or mantras containing specific word sounds can evoke a felt sense of a fundamental sound – its inner sound counterpart. A more direct experience of fundamental sounds is possible only if we utter its corresponding word sound mutely – for example silently mouthing and miming an extended Aaah or Mmm sound. Mouthing and miming specific word sounds in this way we can actually hear ourselves uttering them inwardly – hear them as inner sounds. But the ear with which we hear an inwardly uttered sound is our inner ear, and the voice with which we utter it is our inner voice. It is with this inwardly heard and modulated voice that we can also vary the tone and volume of an inwardly uttered word sound such as an Aaah or Mmm, making it higher or lower in pitch, louder or softer in amplitude, longer or shorter in duration etc.

The inner voice is not something we are normally aware of except when we engage in internal dialogues with ourselves, hearing our own thoughts as mental words, or even uttering them inwardly (as, for example, when we silently swear at someone or something). But even here we are dealing with inwardly uttered words rather than specific word sounds. And even silent voiced word sounds are not themselves fundamental sounds. The closest we come in everyday life to experiencing fundamental sounds is through the felt bodily sense associated with wordless sounds – for example a sigh of relief, a groan of pleasure, an exclamation of surprise or delight, a murmur of satisfaction or a hiss of sarcasm. Fundamental Sounds, whilst echoed in sounds we make with our bodies, are not so much uttered as bodied in the same way that relief may be bodied in a sigh or pleasure in a groan. They are themselves bodyings of felt sense — felt tones and textures of bodily self-awareness. Like their audible expression in sighs and groans they can have different emotional colourations (like a sigh of relief or of resignation). Specific word sounds such as /m/ are just as much an echo of felt sense – of felt tones and textures of bodily self-awareness – as wordless sounds such as sighs and groans, murmurs and hisses. But by their very nature, feeling tones and textures, like musical tones and timbres, can have many different emotional colourations. Every feeling tone or texture can be emotionally experienced and expressed in different ways.

Felt tones and textures, densities and intensities of awareness are the source of emotional experience and expression but quite distinct from them. Similarly, fundamental sounds are the source of speech sounds but distinct from them. Word sounds are uttered with our bodies, giving form to vocal tones. Fundamental sounds are silently bodied, giving form to feeling tones. They find audible expression in wordless sounds such as sighs and groans. But they can also be bodied silently – as silent gestures and movements, looks and facial expressions. They are the felt senses or inner resonances linking not words, but things and people, resonances which take shape as formed expressions of feeling tones and textures, in the same way that word sounds take shape as formed expression of vocal tones and timbres. A hard or soft, rough or smooth, elastic or brittle tone and texture of feeling can be expressed in speech as a hard or soft, rough or smooth, elastic or brittle word sound. But as a fundamental sound, it can also be silently embodied – given silent physiognomic and gestural form.

A letter is the silent face of an utterable word sound. Similarly, facial expressions and other physiognomic forms are silent faces of inner sound – fundamental sounds. Fundamental sounds withdraw into silence in the very act of becoming audible, appearing instead as silent forms and movements – whether these be letters, gestures, the forms and movements of natural objects and living organisms, or the silent shapes of man-made objects. The mystic is someone who can follow sounds into silence, who can listen into silence – and who can hear all apparently silent forms and movements as sounds. Reading involves more than just seeing words on a page. Reading involves listening to and hearing their forms. The mystic is someone able to read in a far broader and deeper sense – to read nature and read human beings as well as letters and words. The mystic does this through an attunement to a 'zero-point' or 'still-point' of inner silence within them that allows them to be touched by the inner sounds or resonances of words and things, objects and events, places and people – their resonant field. The mystic appreciates that not only every letter of the alphabet but every-thing and every-one is the expression of its own fundamental sound. The mystic is someone grounded in their own fundamental tone, aware of all its harmonics, and able to use them as wavelengths of attunement to different aspects of themselves and others. The mystic may or may not be able to sing, but their inner voice has a range of inner tonalities stretching far beyond that of any singer. The mystic can communicate to others directly with this inner voice, change the sound shape and transmitted resonance of their own inner body or organism. For the organism, any organism, is a sound — a shaped psychic envelope of feeling tone. Through resonance the mystic can allow their own organism to take on a shape that is isomorphic with that of other beings. The shape of the mystic's organism is one that they can shift at any time through the use of inner sound and feeling tone. Were one able to behold it visually, it would be seen to freely metamorphose or 'morph', revealing the countless inner faces and forms latent within the mystic's own soul or field of awareness. The mystic is someone able to attune to their own larger identity – not an 'inner self' that is part of them, but a self-field and field-self of which they know themselves to be a part of. The mystic can enter the world of this intensional self – the world of intensional reality, made up of intensional reality fields or planes.

The mystic knows that this intensional reality has a fundamentally musical character. The true mystic is therefore fundamentally a musician, just as every true and great musician is fundamentally a mystic. Being a musician means knowing that sound has its source in silence, that silence leads into a world of inner sound and feeling tone, that all music has its source in this world, and that no audible music could be made, let alone listened to as external sound, were it not for the composer's capacity for inward listening. The mystic is someone in touch and in tune not only with the music of the inner universe and the inner self, but also with the inner music of the outer universe and the outer self, however dissonant or discordant they may seem to be. The mystic is someone who knows as Heraclitus did that "from tones at variance comes perfect harmony" – that harmony and disharmony, dissonance and resonance go hand in hand, and that "the hidden harmony is better than the obvious one." For the mystic, as for Heraclitus, the logos of the *psyche*, is not essentially a type of speech, let alone a verbal account or report but a 'report' in quite a different sense – a reverberation or resonance. "You shall not know the limits of the psyche, no matter how far you go about it, so deep is its inner reverberation." The psyche is the resonant chamber of the human organism, a chamber with a boundless interiority. That is why and how the mystic follows the maxim of Heraclitus: "Listen not to me but to the logos" even whilst knowing that like his own words, this is a logos that men "ever fail to comprehend, both before hearing it, and once they have heard."

How can we possibly comprehend something before hearing it? Only through a mystical and musical appreciation of fundamental sounds – which are not actual sounds or words but pre-sounds or potential sounds. As such they are not less but infinitely more rich in meaning and intensity than any words or audible sounds. Inner comprehension is a resonance with these pre-sounds or potential sounds. Without this inner resonance, we cannot comprehend what is uttered neither before nor after hearing it.

Fundamental Science and M-Theory

When will our physicists realize that their theoretical models are not simply mathematical representations of the outer physical universe of extensional space-time? They are mental metaphors of an inner psychical universe: a non-extensional or intensional universe made up not of fields of energy but of patterned fields of awareness. When will they realize that physical-scientific concepts of fields, particles and waves, or more recently of 'strings' or 'branes', arise within their own field of awareness as mental images – imaginative expressions of particular figurations or field-patterns of awareness?

No better example of the metaphorical nature of scientific models can be found than in Membrane Theory or M-theory: the new so-called 'Mother of all theories' that challenges the 'standard model' of cosmology, looks back beyond the Big Bang and promises the unification of General Relativity and Quantum Field Theory. M-theory had its roots – not surprising from the point of view of Fundamental Science – in a quasi-musical model of the universe called String Theory. In this model, elementary particles are conceived as different excitational modes or 'notes' of vibrating strings, rather like musical strings except with their own intrinsic tension. Strings could be closed loops or open. Stretched across the time dimension they would form tubes or planes respectively. In M-theory, strings can take many forms. A membrane is a two-dimensional string, called a 2-brane. But other types of brane are postulated too, including both 5-branes and 0-branes. In string theory, strings have the peculiar property of being able to 'curl' and 'wrap around space'. They vibrate in 11 dimensions that include three ordinary dimensions of extensional space, seven further spatial dimensions and one dimension of time. In M-theory, every point in extensional space can be described mathematically as a 'curled up' space of seven dimensions. We see in the language of string and membrane theory the most elaborate attempt so far to conceive and picture intensional or non-extensional reality in extensional terms – using both mathematics of higher-dimensional spaces and figurative spatial metaphors such as 'strings', 'membranes' or 'curled up' space.

M-theory is allowing physicists to once again speculate mathematically on the existence of multiple parallel universes, each constituted by a particular membrane shape 'floating' in the 5th dimension. Cosmologists now speculate that our own universe might be the result, not of a primordial Big Bang but of a 'Big Crunch' – the collision of two pre-existing branes in the fifth dimension which brings about a local collapse of that dimension. The uneven distribution of matter in the cosmos is thought of as resulting from the uneven and undulating surfaces of the colliding branes, which are visualized in three dimensions as having different shapes such as spheroids and doughnuts or toroids. The advantage of the theory for physics is that it gives gravity a central role as the fundamental medium coupling matter in different universes of branes. Gravity ceases to be merely an awkward and oddball force within the cosmos we know but the medium linking universes in the 5th dimension. String theory describes dimensions of microcosmic space. M-theory couples this microcosmic but multi-dimensional space with macrocosmic events and the process of cosmogenesis itself. For according to the principle of T-duality, physics in a space time dimension of radius R is equivalent to physics in a space-time dimension of $1/R$.

The language of the new, non-standard model of cosmogenesis based on a mathematics of 11 dimensions is extraordinary in its naïve but duplicitous use of spatial metaphors derived from extensional, three-dimensional space as we know it. Thus we read of branes 'colliding' like particles with one another, 'bouncing' or 'peeling' off one another, or 'melding' with other branes. There is no better example than M-theory of a basic paradox of physical-science – the paradox being that the closer it comes to an adequate and comprehensive model of the extensional universe, the more it (a) relies on ever-more complex mathematics to avoid recognizing the fundamental nature of intensional reality, and (b) unwittingly comes with new and imaginative metaphors such as 'membranes' which are indeed far more adequate metaphors of intensional reality and the psychical universe than previous models of extensional reality and the physical universe.

Concepts such as '0-branes' or 'Calabi-Yau' spaces of seven dimensions 'wrapped' or 'curled up' in every point of three dimensional space are as mind-boggling for the physicist as for the layman. Why? Because whilst they clearly point to the nature of intensional reality and intensional space they do so only in an abstract mathematical way. Were we however, to attempt a fundamental psychoanalysis of the basic metaphors of string theory and M-theory, we might say that the concept of the 'string' is a direct metaphor of a basic dimension of intensional reality – namely thought itself.

The 'strings' of the physicists are a metaphor of their own thoughts, capable of taking different forms according to their excitation mode. Similarly, their 'membranes' are metaphors of their own intensional body or organism with its encapsulating envelope or membrane of awareness. Such an analysis of scientific concepts is of course heretical, for it points to a fundamental deficiency in the way the significance of these concepts is understood. As far as the physicist is concerned, a concept like 'wave', 'field', 'string' or 'brane' has one and only one vector of signification – its physical signification in denoting something 'out there' in extensional reality. The physicists cannot deny that their own imaginative scientific concepts arise from and within their own field of awareness. Yet they can and do continue to deny the psychical signification of these concepts – their meaning as figurative expressions of intensional or psychical reality. Fundamental Science understands the mathematician's figures, the figurative pictures of the physicists and the figures of speech they imply as figurations of awareness giving direct expression to inner fields, field-shapes and field-dynamics of awareness.

The current scientific understanding of M-theory in particular, however, is fundamentally deficient in another respect too. For it not only fails to recognize the psychical as well as physical signification of the concept of 'branes', it lacks any fundamental concept of what a membrane as such essentially is. The mathematical models of M-theory focus on various possible interactions between the boundary surfaces of different membranes. Such models treat membranes as things in themselves, like bodies in space which then happen to interact in a particular way. But the fundamental nature of any membrane is that it is not a 'thing in itself' which interacts with other things at its boundary. Rather the essence of a membrane, understood dialectically, lies in being a dynamic interface or boundary state of interaction. Fundamental concepts, as Hegel recognized, are intrinsically dialectical – they refer not to pre-given elements 'in' a certain relation to one another but to elements of a dynamic relation. Relationality itself inherently is not something static or unilateral but 'dialectical' – dynamic and reciprocal. Frontiers, for example, are not boundaries between separate territories. They are what define those territories in the first place, allowing a dynamic flow of goods and people between them.

In general, a determinate phenomenon such as a cell or any figure such as a circle only ever emerges or stands out (ex-ists) in relation to the background field that is its context of appearance. We cannot logically posit the phenomenon as an entity (+A) without reference to the co-defining field or context of appearance that constitutes all that it is not (–A). Text and context, phenomenon and field, are circularly self-related, defined in and through their relation to one another. (+A) and (–A) are mutually defined elements of their relation, of a singular boundary state (!A). Just as it is a line or frontier that defines the territories on either side of it, so does a circle or three-dimensional membrane of any shape define its own inside and outside. What appears as a figure such as a circle, sphere or surface membrane of any sort is just as much an inner surface and inner boundary of the area or field around it as the outer surface and boundary of the area of field within it. Any two 'membranes' are not just related externally through their possible contact or 'collision' with one another in the 'empty' space or field between them. To think so is to see each membrane only as a surface boundary of the space inside it. If instead we see both circles as internal surfaces or boundaries of the space around them the picture is quite different. For then logically, they are already intrinsically or internally related, simply by virtue of being internal boundaries of a common field or space.

Fundamental Physics as Sembrane Theory

Much attention is paid in our culture to what scientists think, very little to what doing science involves. We think only of physical laboratory instruments and experiments and not of the mental models and metaphors in which the scientist's awareness dwells and through which it finds expression. These constitute a singular and highly specific network of verbal signifiers which constitute the scientist's own mental universe or semiosphere. The information provided by physical instruments constitutes another semiosphere, shaped by the selective significance attached by the scientist to particular signs and signals. The physicists think they are studying and describing invisible and complex features of the outer universe, when in fact they are projecting features of the inner universe or mental semiosphere in which they dwell onto the outer universe.

The surface of a semiosphere is a semiotic membrane or 'sembrane'. Sembrane theory is the counterpart to string theory and M-theory in Fundamental Science. It allows us to understand science, scientists and their theories in a new way – as different universes of discourse or semiospheres. Each of these semiospheres or universes of discourse both in-forms and gives form to the inner universe of awareness that constitutes the scientist's own *noosphere* or sphere of awareness (Greek *noos*). But Sembrane Theory is not just a fundamental semiotics. It is the semiotic foundation of Fundamental Science as such. The notion of the semiosphere transcends the dichotomies of word and world, linguistic and extra-linguistic reality. We dwell within the world as we dwell within the word, both constituting distinct but inseparable semiospheres. There are as many worlds as there are semiospheres – each sphere constituting the surface of a distinct layer of meaning.

The basis of sembrane theory is the semiotics of the French linguist Saussure, for it was he who first introduced the idea of surface fields or 'planes' of signification, using the analogy of a sheet of paper:

"A language can be compared to a sheet of paper. Thought is one side of the sheet and sound the reverse side. Just as it is impossible to take a pair of scissors and cut one side of the paper without at the same time cutting the other, so it is impossible in a language to isolate sound from thought, or thought from sound. To separate the two for theoretical purposes takes us either into pure psychology or pure phonetics, not linguistics."

"Linguistics, then, operates along this margin, where sound and thought meet. The contact between them gives rise to a form, not a substance."

For Saussure a sign consists of two elements, a signifying sound and a signified thought or concept. Signifier and signified, or sound and thought are thus conceived as distinct but inseparable sides of a singular plane or surface boundary. Similarly, specific signifiers and their signifieds can be thought of as distinct but contiguous areas inscribed or cut into this plane, like interlocking pieces of a jigsaw puzzle. The meaning of any given sign is like the boundary contour of one of these areas or puzzle pieces, a contour which both defines and is defined by the contours of continuous pieces – other signs in the language or sign system.

Since Chomsky introduced the idea of language having both a surface and a deep structure, however, no one has thought to relate this to Saussure's model of language. This is unfortunate, for in parallel with M-theory, we can conceive not only of two-dimensional planes or membranes of signification — 'sembranes'. We can also conceive of three-dimensional ones. Indeed Saussure himself implies a notion of semiotic depth by introducing a picture of wavy regions or fields of amorphous thought and sound bounded by the signifying plane or sembrane.

"So we can envisage...language... as a series of adjoining subdivisions simultaneously imprinted both on the plane of vague amorphous thought (A) and on the equally featureless plane of sound (B)...what happens is neither a transformation of thought into matter, nor a transformation of sounds into ideas. What takes place, is a somewhat mysterious process by which 'thought-sound' evolves divisions, and a language takes shape with its linguistic units in between those two amorphous masses."

His analogy is a telling one: "One might think of it as being like air in contact with water: changes in atmospheric pressure break up the surface of the water into a series of divisions, i.e. waves. The correlation of thought and sound, and the union of the two, is like that." For what this picture also suggests is the possibility of a deep oceanic undercurrent of thought – a deep structure – more or less manifest in the surface waves or undulations of the sembrane.

Lacan adopted Saussure's ideas of chains of signification – analogous to the strings of physics — linking one signifier with another, but denied any fixed relation between signifier and signified. For him, planes of signification were composed of signifiers only. On the one hand the signified latter could 'slide' under the signifier, like undercurrents in a sea. On the other hand no signified could be identified as a thing in itself, but rather constituted a 'no-thing' that could manifest only as signifier – a wave or wave packet on the plane of signification.

Fundamental Semiotics, as a multi-dimensional field semiotics, understands planes of signification as sembranes surrounding or bounding an inner semiotic space with a depth dimension of meaning or sense, and consisting of intrinsically meaningful flows and streamings of awareness as such. From this point of view the very language of String Theory and M-Theory, like the languages of other sciences, are themselves semiospheres – sembranes of interrelated signifiers, surrounding a *noosphere* or inner semiotic space of awareness (*noos*). The mathematical problems involved in describing the spatio-temporal dimensionality of strings and branes in physics is an expression of its failure to acknowledge the non-extension or intensional nature of semiotic space or meaning space — the noosphere enclosed or bounded by semiospheres and sembranes. What physics calls the fifth dimension is no dimension of extensional space-time but is the depth dimension of meaning or sense that constitutes the inwardness of any word or signifier – its intensional or semiotic space.

Viewed within our perceptual semiosphere, a book is a bounded three-dimensional body in extensional space. Each of its pages is a two-dimensional sembrane. As soon as we start reading the book however, our awareness flows into its interior semiotic space – its noosphere. Physically, the book remains the same size. But feeling our way into its concepts or characters, ideas or images, its internal space of meaning expands. As our own inner awareness flows into it, we experience an inward expansion of awareness within the book – not as a physical phenomenon but as the visible outer surface of a multi-dimensional world of meaning – a semiosphere with its own internal noosphere. We experience different layers of meaning within it, each of which constitutes a semiosphere of its own. The mental associations and images that flicker in our mind as we read a novel for example, represent one such semiosphere. The felt bodily sensations and emotions it evokes are another.

Dwelling within the intensional meaning space of the book we lose consciousness of the book as an object in extensional space. It is as if a singular noospheric space of awareness within our own bodies and minds merges with the internal noosphere of the book.

According to Derrida, however the "presumed interiority of meaning is already worked upon by its exteriority". This is true, but then again there is no such thing as an exteriority without an interiority. The fact that any sembrane or semiosphere consists of a network of mutually related signifiers does not imply that there is no interior semantic depth dimension to these planes or spheres of signification. Fundamental Semiotics is not a semiotic reductionism which treats sense as a product of the signifiers and the symbolic. It recognizes however, that sounds and other signifiers not only articulate or express sense but attract and give form to sense. Fundamental 'sense' however, is not signified sense but felt sense – it consists of intrinsically meaningful field-qualities and field–patterns of awareness that constitute the inner noospheric dimension of any semiosphere. The fundamental, felt sense of any verbal or perceptual signifier is reducible neither to its relation to other signifiers in a plane or sphere of signification, nor to its reference to a 'signified' – a pre-given meaning or thing.

Thus the sense of a sentence is not the product of a formal syntactic structure of signification built up upon separable elements such as nouns and verbs, determiners and modifiers. It is the other way round. A unitary sense manifests itself as a syntactic structure or field pattern of signification. This unitary sense, unlike the sentence, is indivisible – it cannot be broken down into individual components. It is a coordinate point within a semiosphere.

Strings of letters or words constitute signifiers only because they share a common *coordinate point* in semiotic space. Similarly, multi-word lexical units as such as 'fridge freezer' or 'box of matches' or 'on top of the world' are rightly called 'collocations' because their individual components, as elements of this unit, share a common locus or coordinate point in semiotic space.

The relation of semiosphere to noosphere, significance to sense, is a relation of surface strings or planes of signification (1-sembranes and 2-sembranes) to coordinate points within semiospheres or 3-sembrances. Each of these coordinate points is in turn the semiotic equivalent of a 0-brane in M-theory. Coordinate points in intensional space are centres not only of felt sense but intent. When we attend to the inner resonance of a word, phrase or sentence we attune ourselves to the coordinate point of that signifier in semiotic space. When we intend a sense we activate such a coordinate point, attracting potential signifiers and strings of signifiers in resonance with its basic tonality.

Intent is an attunement to potential field patterns of signification, actualised through the activation of coordinate points in intensional space. These coordinate points manifest not just in patterned strings of words or sentences, but in patterned perceptual phenomena and sequences of events. Each coordinate point in intensional space has counterparts in extensional space and in the human body itself. When we describe someone as 'talking off the top of their head' or 'speaking from the heart' or 'expressing a gut feeling' we are referring to the locus or coordinate point of their intent.

People need to make sense of their lives, and do so in very different ways. If they have difficulty in making sense of their lives they may signify this difficulty in different ways too – through behavioural signs or through bodily symptoms of one sort or another. If these get out of hand, they may also seek help from a friend, counsellor or therapist in 'sorting themselves out' i.e. making sense of their lived experience of themselves and of the world. For most people, including many counsellors and therapists, however, 'making sense' of experience means finding new ways to signify it – whether verbally or non-verbally, through words or body language. A counsellor for example, may perceive the signs of emotion behind a person's words, and help a client to signify their emotions more precisely or more fully – either in words or through body language. But signifying something we feel is not the same as feeling its significance. Signifying feelings keeps us trapped in the world of conventional signification. The opposite to signifying what we feel is to feel the sense of particular signifiers. The opposite to signifying what we sense is to sense its significance. Feeling significance leads us into the world of sentience. This can be defined either as felt sense or as sensed significance or 'sentience'.

If someone talks about an event, emotion or experience of any sort that has meaning for them, then they are signifying this meaning or sense. They may signify this meaning with words or signals conveyed through their body language. But what is signified is not the experience they describe but the sensed meaning or significance it holds for them. The experience itself, the event or emotion for example, is just as much a signifier of this sensed meaning or significance as the words and body language with which they signify it.

Consensual reality is a linguistically and perceptually pre-structured world of signified sense – a semiosphere of conventionalised signs and sign systems. But behind and within the world of consensus reality is another world. This is not a semiosphere of physical or verbal signs – of signified sense. Instead it is a noosphere of sensed significance or sentience.

The 12th Fundamental Distinction

The twelfth fundamental distinction is the distinction between signified sense – consensual and conventional signification, and sensed significance or sentience. A semiosphere is a semiotic membrane whose outer surface is a plane of signification or signified sense and whose inner surface is a plane of sensed significance. No word or thing can have meaning as a sign save as a phenomenon in the field of awareness (*noos*) of a sentient being. Sentience is the condition for semiosis – the signification of sense and the sensing of significance.

Conventional sign systems and ordinary 'con-sciousness' create a con-sensual reality in which common structures of language and perception overlay sentience and blind us to simference. In doing so they remove from our semiosphere, from the signs all around us, what Martin Buber called the 'seed of address'. This is the capacity of a word or event, person or thing to alter and expand our very sense of self through awareness of simference. Sentience is sensed significance. Sensed significance is sensed simference. If followed, this sensed simference will on each occasion transform our felt sense of self and open us to new potentialities of awareness and action, new modes of self-experience and self-expression, and new dimensions of language and perception.

The relational meaning of signs or signifiers themselves is not reducible, as both Derrida and Lacan would have it, to their differences but to their simferences. The identification of meaning with difference alone has its roots in Saussure. According to his model of language, the meaning of the word cat is a function of its difference from other words mat, hat etc. There is no intrinsic relation between the sounds of the word cat and its signification. For Saussure, a fundamental semiotic principle of linguistics is the arbitrary and purely conventional nature of the signifier, which derives its meaning only through its difference from other signifiers in a patterned plane of signification. In contrast, Martin Buber stressed the non-arbitrary nature of the sign – for even conventional signs possess, in the living context in which they address us, a sense which is always unique and unrepeatable – irreducible to other signs and indefinable by them.

Buber's semiotics is a dialogical semiotics in the true sense, based on the understanding that what communicates through the word (*dia-logos*) is an essentially wordless sense which cannot itself be signified in words. Words do not represent but convey senses, senses that cannot by nature be represented in other words. Whenever we interpret and represent, define or describe the sense of a word with other words, these words will instead convey a sense that is uniquely their own. Any interpretation by which we seek to represent the meaning of another person's words conveys its own message – the representation of a message is always a more or less disguised response to that message. What is true of words is true of all signs. The sense of a sign is something unique and unrepeatable. It is not a property of the sign and cannot be represented in signs. Instead it reaches out to us through the sign. Whether this sense reaches us depends on our own sentience.

A sign is not essentially a sign of something – whether a thing or another sign. It is indeed a vehicle, the bearer of a felt sense beyond signification. Sense is not the property of signs. Signs are a vehicle of sense. The fundamental sense of a sign is no thing and no word, nothing we can look up in a dictionary or point to in the world. Unlike signs, senses are nothing that we are aware of. Instead they are felt dimensions and directions of awareness as such – subtle tones and textures of awareness, subtle qualities and intensities of awareness, subtle configurations or shapes of awareness, or subtle movements, flows or shifts of awareness from one dimension or direction of awareness to another. Sense is no signifier and no signified, no word and no thing that we think of. It is a felt relation to something or someone other than self. When a person's words or body language convey a 'sense' of warmth for example, what we 'sense' is not merely something we are aware of in the other. We bask in the warm quality of their awareness of us. Awareness and its qualities are intrinsically relational. The warmth we feel flowing from the other is a sense of their own aware and felt relation to us. Awareness is also intrinsically reciprocal. We could not feel the warm glow of another person's awareness of us if our awareness of them was not being warmed by it, reciprocally altering our felt relation to them.

Fundamental Science and Semiodynamics

Fundamental Science recognizes an intrinsically semiotic dimension to the universe and its fundamental dynamics. For the relation between fields of awareness and the phenomena that manifest with them is at the same time a relation of inner meaning sense and its outward signs – and of sentience or sensed significance to signification or signified sense.

- Sense prefigures the emergence of signs, consisting of potential patterns of signification pregnant in felt dimensions and directions of awareness but not yet manifest as phenomena – as verbal or perceptual signifiers.

- Signification is the conversion of sense into signifiers – into phenomena we are aware of. Sentience is the reconversion of signifiers into sense – into felt dimensions and directions of awareness.

- *Phusis* or 'emergence' is also *semiosis* – the energetic materialization of sense or meaning in the material word, and its reconversion into sense through sentient awareness of the material world.

- Sense is not a relation of signs to one another or their reference to a pre-given object.

- The fundamental relation of sign and sense is one of resonance – a resonance between potential and actualized field patterns of signification.

- 'Energy' is the formative activity of signification, translating potential patterns of signification, latent within fields of awareness, into outward phenomena — verbal and perceptual signifiers.

- The primordial ground state of awareness is a field of unsignified sense and unsensed significance that is the energetic source of all signified sense and sensed significance, all consciousness and all sentience.

Fundamental Thinking

When we think of science we think of laboratories, instruments and men in white coats. We take for granted what is most obvious, namely that science is first and foremost a semiotic activity – dealing with patterns of significance – and an activity of thought. To think anew about the fundamental nature of science, therefore means to think anew about the nature of scientific thinking as such. Like ordinary thinking, scientific thinking takes for granted that its subject matter is things 'out there'. 'Thinking' is understood as having thoughts about things, representing the relations between these things and in this way providing an account of their nature. The basis of scientific thinking is direct sensory perception of those things or a type of indirect instrumental perception. Its purpose is explicitly semiotic – to 'make sense' of things. This is understood as finding some sort of rational order in the universe that can be represented in formal signifiers – whether mathematical symbols or scientific concepts.

The term 'making sense' has another sense however. That is to find meaning or sense in them, rather than constructing a representation of that sense. We assume all too easily that these two senses are the same – that finding meaning in things is to find a rational order of the type we can represent in sign systems – in signifiers of one sort or another. One basic problem of scientific thinking is that sign systems such as language already possess an order of their own. Another is that the 'things' that scientific thinking investigates are themselves shaped by these sign systems – they are themselves signifiers. In a quite general sense, our very physical perception of things itself is something already informed by verbal concepts which bestow on them a particular significance. We hear a sound as a 'car' passing by. We perceive a visual shape as a 'clock' on a mantelpiece. 'Car' and 'clock' are not pre-given objects of sensory perception that we then name or signify in words that are part of our linguistic vocabulary. 'Car' is already a type of 'word' in a perceptual vocabulary. To begin with, the infant does not hear or see 'cars' or 'clocks'. It hears sounds and sees shapes. It does not hear these sounds or see these shapes as pre-conceived 'things' at all. What as adults we take for granted as pre-given objects of sensory perception are perceptualised concepts already imbued with a particular significance.

This being the case, how can scientific thinking escape the vicious circle of simply representing in concepts perceptions, sensory or instrumental, already in-formed by those concepts? The answer seems clear – it can do so because the perceived relationships of things do not always match their conceptual or mathematical relationships represented in sign systems of science – in its theoretical models. To question a theoretical model is one thing however. To question the nature of the very 'things' whose relationships it is designed to represent is quite another. To question a model of the relationship between matter and energy, for example, is quite a different matter from questioning what the things we call 'matter' and 'energy' essentially are. This type of questioning is more fundamental and likewise the type of thinking that it requires and the type of 'science' it produces.

The semiotics of conventional scientific thinking is essentially literalistic in character – it assumes a one-to-one relationship between its signifiers and the things they signify. What it essentially explores, however, is the relationship between two sets of signifiers or two sign systems – those that constitute its own system of formal linguistic and mathematical signifiers and those that constitute perceived reality. Its object is to match up these two sign systems – the system of formal signifiers represented by its concepts and mathematical signs and the system of physical signifiers represented by its percepts and instrumental measurements. At the same time however, it identifies the system of physical signifiers with reality or empirical fact.

Science seeks to grasp the 'factual' truth of things in thought. It reduces the meaning of things to their perceived form and behaviour – a form already imbued with conceptual significance. Ultimately therefore, it makes sense of the universe only as a set of mutually related physical signifiers or mutually defined formal signifiers. But there is another way of 'making sense' of the universe – one that does not reduce 'truth' to a correspondence between scientific models and empirical evidence, theories and 'fact', formal and physical signifiers. This other way of making sense of the universe is based not on sensory perception alone, informed as it is by pre-conceived patterns of significance, but on the felt meaning or significance of things – on what Eugene Gendlin has called 'felt sense'.

If someone describes another person as a 'dark horse' we do no take their statement literally, but 'metaphorically'. That is to say, we have a felt understanding of what they mean. We do not conduct scientific experiments and look for 'evidence' that might prove whether this person is 'really' a horse and a dark one at that. We would be quite wrong in thinking that this is common sense with which science has no dispute. 'Scientific' psychiatry, for example might well lead a professional psychiatrist to perceive a patient not merely as a 'dark horse' but as 'psychotic'. Why? Because this patient claims, for example to have a bomb in their belly. From the scientific point of view there is no 'empirical evidence' whatsoever for this claim. Therefore it is irrational and a sign of psychosis. The same individual who would be quite ready as a human being to accept a description of a friend or colleague as a 'dark horse' may, as a professional psychiatrist, be totally unwilling to treat the patient's claim as anything but an assertion of literal fact. Both are unwilling or unable to take the assertion as the metaphorical expression of a felt meaning. Psychiatrist and patient, in other words, share a common mental disorder – the disorder of literalism. The patient claims it to be a literal 'fact' that there is a bomb in his stomach. The psychiatrist not only accepts the literality of the claim but takes it as a sign that the patient is 'psychotic' – and believes that there is quite literally, some 'thing' inside him (a chemical imbalance in the brain for example) that is 'causing' this psychosis. This 'scientific' attitude is adopted even though there is no more evidence of a chemical imbalance than there is of the bomb. Both psychiatrist and patient are mad – they share the common mental disorder of semiotic literalism – a type of semiopsychosis accepted as normative and actively cultivated by conventional scientific thinking.

One reason the patient finds it necessary to insist that there 'really' is a bomb in his stomach is that in a scientific culture facts and their significance are treated as more real and taken more seriously than feelings and felt meanings. The reduction of felt meaning to its expression in verbal signifiers and assertions has a profoundly distorting impact on our understanding of what thought is. Scientific research revolves entirely around the relation of verbal signifiers and assertions to literal facts rather than to felt sense. Quantitative statistical 'significance' replaces qualitative felt meaning or significance.

Literalistic thinking does not recognize that perception itself has an intrinsically metaphorical character. Aristotle defined metaphor as the verbal representation of one thing as another – describing a person as a 'dark horse' for example. But our very perception of things and people may have a metaphorical character in a deeper sense. This deeper level of metaphor lies in perceiving something or someone 'as' this or 'as' that in the first place – as a 'car' or 'clock', as a 'bomb' or as a 'brain imbalance', as a 'psychotic' or as 'a dark horse'.

Plato recognized long ago that the perceived form of something is no 'thing' in itself – it has no mass or energy. We may perceive something as red or round, as a tree or as a horse, but we cannot pick up or measure redness or roundness as such, nor can we climb the treeness of a tree or ride the horseness of a horse. We can measure and compare the diameter of two round objects or ascertain their degree of roundness, but not roundness as such. The perceived form of a thing, 'look' or 'aspect', is what the Greeks called *eidos* from whence we gain the word 'idea'.

The word 'aspect' and 'look' have themselves a double sense or aspect. On the one hand they refer to an 'objectively' perceived face or facet of something, for example a facial expression or the facade of a house. On the other hand, to a 'subjective' angle or perception. Aspects are in this sense something more fundamental than either 'subjects' or 'objects' of perception, constituted as they are by a relationship between a look or face of something or someone, and a way of looking at it. This was something recognized by Winnicott, who observed that when an infant looks into its mother's eyes, how they look is an expression of her way of looking at and seeing the infant itself. More than this, it is an expression of her felt resonance – or lack of it – with the look on the baby's face. A look on someone's face is not only an object of observation. It reveals their gaze – their way of looking out at and seeing the world and other people. Winnicott noted how if the mother is not in resonance with her baby, her face will express only her own moods and preoccupations. As a result the infant can find no reflection of itself in her look, nor will their mutual gaze constitute a form of resonant contact and communication. All the infant can do then is attempt to study her face as an object. Like a meteorologist 'reading' the weather in order to predict its patterns the infant must 'read' the signs on the mother's face in order to gauge her mood and predict her behaviour. The baby becomes a precocious scientist in other words whose way of looking at and 'reading' physical signifiers has an analytic and predictive function, rather than being an expression of resonance with the qualities of awareness that shine forth through the look.

Physical signifiers do not become phenomena in the fundamental Greek sense of the word – that which shows itself or shines forth without needing to be turned into an object. But the moment we turn another person's eyes into a mere object of visual scrutiny, we cease to perceive their look as a way of looking, we cease to perceive their gaze or the inner light it radiates.

As Martin Buber put it: "The child that calls to his mother and the child that watches his mother – or to give a more exact example, the child that silently speaks to his mother through nothing other than looking into her eyes, and the same child that looks at something on the mother as at any other object – show the twofoldness in which man stands and remains standing".

For the psychiatrist confronted with the patient who believes there is a bomb in his belly, the relation between this 'sign' and the patient's 'psychosis', is merely an objective diagnostic 'reading' – quite independent of the psychiatrist's relation to the patient. In fact, the psychiatrist's 'reading' of the sign is the expression of a very particular mode of relation which Buber called the "I-It" relation. The patient's signifier – the bomb – is not read as the phenomenal revelation of something felt by a human being. Like the physician, the psychiatrist reduces phenomenal signifiers or physical signifiers – mere bodily or behavioural 'signs' of 'It' – to some 'thing' in the form of a chemical imbalance in the brain. It is not understood through felt sense but only through a medical–scientific signifier such as 'psychosis'.

When we 'see' that somebody is unhappy or tense, or 'hear' frustration in their voice this is not a deduction made from some sort of quasi-clinical observation. Similarly, if somebody looks or sounds 'unwell' to us we are not, like the physician, medically interpreting certain overt diagnostic 'signs'. What Heidegger refers to as 'genuine' seeing or hearing is not in the first place a seeing or hearing which has as its object something or 'some-body' in the literal sense, but rather 'some-one' – a being and not a body in space and time. What Martin Buber called the "I-Thou" relation is a direct intensional relationship to another being rather than an objectifying perception of an extensional body – one that brings with it a quite different perception of that body and a quite different 'reading' of physical phenomena in general.

"Man is encased in an armour whose task is to ward off signs…What occurs to me addresses me. In what occurs to me the world-happening addresses me. Only by sterilizing it, removing the seed of address from it, can I take what occurs to me as a part of the world-happening which does not refer to me." (Buber) Reading signs in the scientific or medical manner is a way of warding off their felt significance and not having to respond to that felt significance. The understanding of truth as correspondence between ideas is used to avoid 'correspondence' in the deeper sense of echoing responsiveness or resonance – co-responsiveness and co-responsibility for what is.

A thinking response is a response in resonance with what addresses us, expressive of its felt meaning to us as aware beings. Thinking seeks answers to questions. Scientific questions are thought to arise through logical incoherencies, empirical incongruities or lack of completeness in a theoretical model. But a question is fundamentally something quite different. It is a felt dissonance or lack of resonance between felt sense and its signifiers, one that expresses itself in some form of incongruity, incoherence or lack of completeness in a set of signifiers. The signifiers in question may be either formal signifiers such as verbal concepts and mathematical signs or physical signifiers in the form of observational percepts and empirical data.

Plato's recognition that perceived form is essentially immaterial or ideal led him to understand perceived reality as a type of shadow world – an idea construction whose source lies in a realm of pure 'forms' or 'ideas' which only thinking can gain access to. But from a field-phenomenological perspective 'ideality' as such is not a static realm of pure forms of ideas but a dynamic field of formative and transformative activity – of 'energy' in the root sense of *energein*. Thinking as a mental activity is a direct expression of this formative activity and in turn in-forms or patterns it. Both formed concepts and formed percepts are an expression of an active, formative intelligence or energy — one that not only in-forms our thoughts but in-forms things themselves, that not only in-forms our minds but in-forms our own bodies and all bodies.

But thinking in this fundamental sense – as the mental expression of formative activity – is not the same as having thoughts about things. As long as we merely 'have' thoughts about things we are not, fundamentally speaking, thinking at all – for we are not aware of our own thinking as a formative activity. That is why to think in a fundamental way is not merely to have thoughts about particular phenomena that we are aware of. It is to be aware of our own activity of thought as an expression of felt sense – potential field-patterns of significance. Formed products of thought activity – 'thought forms' – in turn actively in-form reality through their morphic resonance with particular patterns, bringing them into actualisation. Fundamental Concepts, however, are concepts which do not simply give form to felt sense but stay in resonance with the larger field of still unmanifest meanings or patterns of significance.

"Method is awareness of the form taken by the inner spontaneous movement of its content." (Hegel). Awareness of the form taken by the spontaneous and dynamic movement of thought is at the same time the basic 'method' by which we transcend not only fixed concepts and conceptual frameworks but also fixed percepts and perceptual frameworks – for the latter are reinforced by resonance with the former. Wherever and to whatever extent we are identified with a thought we cease to be aware of it as a thought but take it as a thing in itself. Wherever and to whatever extent we are identified with a concept or conceptual framework, we cease even to be aware of the concept or conceptual framework but experience it instead as a percept or perceptual framework. We see a tree only as a tree, quite unaware that the 'thing' we are seeing is already in-formed by a thought form or concept. We forget that what we perceive as a tree is but one form, idea, aspect or 'as-spect' of what the tree itself essentially is – one actualized form or field-pattern of manifestation emerging from a field of potential forms or patterns.

When we hear 'a car passing by on the street' our unconscious identification with concepts of cars, streets, trees etc. is so complete that our identification of those cars, streets and other phenomena appears to us as a perceptual process alone. It is only through direct inner awareness of our own thought activity that we transcend unconscious identification with its conceptual products and extend our very perception of reality as a result. The term 'unconscious identification', however, is in one sense both an oxymoron and a contradiction in terms. For 'unconsciousness' as such consists in being identified with a concept.

What Freud called 'unconscious' therefore, is not a container of unconscious thoughts, but consists of all those thoughts we are quite consciously identified with. Being consciously identified with these thoughts however, we cease to be consciously aware of them as thoughts – and as expressions of our own formative, thinking activity. And yet it is only through this lack of consciousness of our own thoughts as things in themselves, through being identified with them, that we are capable of having conscious percepts of things at all – of seeing 'trees' or hearing 'cars' pass by in a 'street'.

Unconsciousness means being passively identified with a concept and experiencing it consciously as a percept instead. Awareness on the other hand, is the ability to actively identify with a concept. For in actively identifying with something we retain an awareness of ourselves as distinct from that thing and therefore not identical with it. Fundamental Thinking, as a direct awareness of thought activity, allows a conscious and active identification with its products – with particular concepts – thus overcoming a state of passive identification with them.

Awareness of concepts allows us to identity them, refine them and form them into new conceptual frameworks. This helps us to disidentify from old concepts and free our perception from them. But Fundamental Thinking is also a capacity to actively identify with new and unfamiliar concepts, thereby actively re-shaping our perception. Dream experience is an inexhaustible source of both new concepts and new percepts. That is because the very process of dreaming puts us in touch with felt sense in a way that is disentangled from many of the 'root concepts' that we are identified with in waking life and which in-form our waking perception of reality. But in dreaming our awareness can also flow into an identification with new concepts in a way that generates new percepts – opening up dimensions of perceptual reality – dimensions that may be more vivid and real – more lucid – than waking reality itself. This happens particularly in so-called 'lucid' dreams in which the dreamer is aware of dreaming. For this conscious awareness of dreaming allows the dreamer to distinguish themselves from the thoughts that take shape as dream percepts and instead to consciously and freely identify with new thoughts.

Being conscious that we are dreaming, we can consciously identify with the thought of flying. As a result we can and do fly, even though the idea of human beings levitating or flying is something inconceivable for most people in waking life. Flying however, is only one of the many otherwise inconceivable possibilities we can not only conceive but realize in dreams, with the necessary degree of consciousness or lucidity. Even without this lucidity however, dream awareness being more fluid than waking awareness, can unconsciously flow into new patterns, allowing us to identify with new concepts and experience them as perceptual realities. Conversely, by virtue of the greater natural mobility of dream awareness, dream experiences can be an inexhaustible source of fundamentally new concepts as yet unconceived in our waking world – but nevertheless fully capable of being realized within it. That is why there is hardly a single major scientific technology fundamental to our waking world that does not have its ultimate source either in daydreaming or in remembered dream images and experiences – the oil industry and its products (the dream inspiration of the chemist Kekule) being but one example. It is those thinkers who are also dreamers, and those dreamers who are also thinkers who have the most fundamental and lasting impact on the waking world. Dream experience is the direct expression and experience of a mode of thought activity richer and more fluid than any conceptual framework of waking thought – expressive not only of already actualised and already experienced field-patterns of awareness but of limitless fields of potential ones.

Waking thought is based on our perception of the outer relationships between things and between people. Dreaming is a direct experience of our inner relationship to things and people. This in turn can give deeper conceptual insight into the inner relationships between them. Fundamental Thinking, as an aware activity of thought in sustained meditative resonance with felt sense, provides a bridge between the worlds of waking and dream reality. It arises from a distinct layer of awareness between these worlds.

Fundamental Identity

The idea of reincarnation is so foreign to the conceptual framework of Western science and religion, because the fundamental nature of identity, and with it, the nature of incarnate existence has not been adequately addressed. Incarnate human existence is viewed as a correlation of one self with one body. The self is seen as a localized subject bounded by the physical body. Science acknowledges that the body we possess now is not even composed of the same matter it had at birth. The very atoms and molecules that constitute our bodies and constitute, in this sense, our material identity, were all once elements of other bodies – mineral, vegetable and animal. They are transmigrating fragments of material identity, united only by their bodily patterning and interrelationships. But the question of when and in what manner the air you breathe or the food you digest becomes you (or ceases to be you) is more than just a question of when and in what manner its material elements become part of your bodily constitution and patterning. For what applies to the material elements of the body applies also to the psychic elements that compose the self. As Seth puts it: "Each identity is itself and no other, and yet it is composed of myriad fragments of other identities." It is these fragment identities that manifest, mix and merge in our dreams, but perhaps not only in our dreams. "All consciousness is interrelated. It flows together in currents, rises and falls, eddies and breaks, mixes and merges."

If identity is fundamentally a pattern of identification, as Seth suggests, then the fundamental nature of identity – of beings – cannot be conceived of as static but hinges on the nature of identification itself as an activity, the activity of be-ing. The psychoanalytic understanding of identity and identification rests on optical metaphors – we 'see' ourselves 'reflected' or 'mirrored' in others or 'project' aspects of ourselves 'onto' others. There is no concept of awareness paralleling our understanding of matter — as something that can flow in currents, that has both a particle and a wave character. Nor is there any concept of the self as something with a non-local or field character, the self-expression of a field of awareness through the actualization of potential field-patterns or gestalts of awareness. A more fundamental understanding of identity would recognize that awareness not only forms itself into patterned figures but flows into and out of these figures, which in turn can affiliate to form more complex patterns or gestalts of awareness. The identity of any being as such an awareness gestalt is not fixed however. Its overall boundary or psychic envelope is permeable in the same way that a cell membrane is, allowing flows of awareness in both directions.

We have no difficulty acknowledging that an actor's awareness can flow in and out of different parts, and that he or she, despite 'having' only a single body, can embody different 'selves' – those aspects of themselves in tune with the parts they play. These aspects consist of different actual or potential patterns of bodily self-expression, of speech and movement, each with their own overall tone or resonance. The expression of these aspects in performing a particular part in turn in-forms those aspects, expanding and enriching them with new elements drawn from others. But once performed, a part becomes easier to play again, having formed up its own enduring field-pattern in the psyche of the actor, and stabilizing itself through morphic resonance with this field pattern. The actor need only attune to this pattern, and allow their awareness to once again flow into it, for them to achieve the level of identification with the part necessary to embody it in performance – to translate it into patterns of speech and movement, gesture and facial expression which convey its overall tone or resonance. So it is, that on a less dramatic scale, we all act out different aspects of ourselves in different situations and with different people, on different stages and with different casts – and with different intensities of involvement, different degrees of identification. It is in this sense that the self we know is indeed a pattern of identifications, or a pattern of patterns – some so familiar or relatively unchanging that we do not consciously and actively identify with them – moving in and out of role in the way an actor does – but instead are passively identified with them. To an extent therefore, our primary identity is not a conscious identity at all but as Freud suggested, an unconscious one. This is not because this identity is buried deep in a hypothetical 'unconscious' but rather the opposite, because we are totally identified with it in our conscious life.

These reflections on the relation between potential form or pattern and embodied per-formance, on conscious acts of identification on the one hand and unconscious patterns or identities on the other, are crucial if we are to understand the fundamental nature of identity, and 'incarnation' and 'reincarnation', in a new way. Actors do not simply make calculated use of their bodies to enact a part. They body that part, allowing aspects of themselves to incarnate through it. As a result these aspects or expressive potentials are both actualized and added to, both given form and transformed in performance. The newly formed field-patterns however, do not disappear or 'die' when the show comes to an end, when they cease to find bodily expression and the stage is abandoned. They endure as stronger, more highly stabilized field-patterns or selves within the field of awareness or self-field of the actor.

The question of whether the self or soul survives the death of the physical body begs the other, more fundamental question –

the nature and identity of this 'self' or 'soul'. From a field-phenomenological perspective, there is no single self or identity in the first place. Instead identity is a dynamic grouping or gestalt of 'aspect selves', each of which in turn is one expression of a larger field of awareness – the word 'soul' being a name for this larger identity or field-self.

It is because all potentialities that we embody and express in a given life are the self-actualization and self-manifestation of this field-self or soul that we can be said to have an identity — a self or soul. But that is not to say that the identity of this self or soul can be reduced to its embodiment or incarnation in any one life – just as it cannot be reduced to its embodiment or incarnation in any one role or persona we adopt in that life. The 'life after life' can be compared to the off-stage life of an actor, whose actual personality can as well be narrower or broader, less or more vivid and colourful, than those parts they have performed – but whose larger identity and soul-life will indeed have been enriched by them, and bear within it the potential to perform many new parts. In our life after 'death', as in our life before birth and as in 'life' as we know it – between birth and death – everything hinges on the relationship between the self as we know it, and the soul – between the self we are identified with and that larger field of possible identities whose self-manifestation we are.

The Western concept of incarnate existence is a myth of singularity – one self, one body, one life, one physical world or universe. The Eastern concept of reincarnation also retains a notion of singularity, with the important difference that the one self can express and experience itself in different ways in many bodies, many lives, many worlds, fields or planes of reality, physical and non-physical. The central difference between them hinges on the nature and identity of this 'one self'- whether it is identified with the personal self we know in this life, with some

195

deeper 'inner self', with God or a god, or whether it is understood in a deeper way: not as any bounded identity, inner or outer, but as a soul in the broader sense – a dynamic and patterned field of identifications. This soul or field-self includes not only potential and actual identifications but past ones that persist as dormant identifications, and future ones that exist as latent identifications – potential identifications with a higher probability of manifestation than others.

Neither the Western nor the Eastern model of incarnation and identity fully acknowledge the fundamental equivalence of incarnation as identification. Understanding this equivalence, the concept of reincarnation does not appear in any way esoteric. For then we see that our lives between birth and death involve a constant process of reincarnation – understood not as the survival into eternity of a single identity but as a constantly shifting pattern or cycle of identifications that centres now on one, now on another identity, role or persona. We are all constantly aware of a field character to our own identity, if not the unbounded reality of our own soul or field-self, then at least of our own self-field – the field of possible identifications that we draw on in every decision we make, no matter how minor. For our every word and deed, our every act, mental or physical, involves at some level an act of identification with a particular pattern of action, a particular identity, one that implies the existence of other possible patterns of action, other possible identifications and identities.

Acts of identification, by their very nature, alter the identity of the agent. Whenever we identify with a particular pattern of action, a part our awareness flows into and fills that pattern, whilst other parts flow into alternate possible patterns of action. The self that identifies with one course of action is not the same as the self that then finds itself identified with that course of action, and knows itself only as the self that came about through that course of action. Nor is it the same as other alternate selves that

196

came about through identification with alternate courses of action. For the self that has chosen a given course of action (X), these alternative selves are purely imaginary – selves that might have done Y or Z. But from the point of view of the selves that chose Y or Z, it is the self that has chosen X that is the 'imaginary' self, the self that might have been. Alternate selves occupy parallel alternate realities. They are possible actualizations of a field–self that is the source of all potential actions. As such they are also connected to one another through this source field and field–self, and aware of one another through their own self-field. The self-field is the very foundation and very fabric of our inner life, for every moment of our lives, in every situation, at every point and in every place, we are aware of pregnant potentialities. We no sooner identify one of these potentialities as a possible course of action (whether a simple action such as getting up to make a cup of tea or divorcing from a partner) that other alternate courses of action become possible at the same time (not getting up or not divorcing). Every possible course of action will affect not only our lives but our very identity, for actualizing it will involve an act of identification. The self that chooses X becomes a different self to the self that chooses Y or Z. The self-field includes past as well as future possibilities. Within it, we retain an awareness of actions we could have undertaken as well as those we could still undertake. And what applies to possible actions applies also to possible events. Our self-field is an awareness both of different events that could yet occur, those could have occurred and those that did occur – but were experienced only by alternate selves in their own alternate realities. It also includes actions and events that remained merely potential – did not become occasions of conscious choice. All actions constitute events in themselves, bring events about, and bring other possible actions and events in their wake.

Fundamental Theology

If the meaning of a word is its soul, irreducible to its physical manifestation as ink marks on a page, pixels on a screen, or sound vibrations in the air, then this soul, like our own, does not dissipate or 'die' when the last sounds leave our lips or the last page of our life is turned. The felt meaning of the word preceded its birth into speech, and lives on in the listener after the speaker passes into silence – to be reborn at other times, and reshaped in other words. Indeed it lives on in the speaker's soul too, for as the Christian mystic Meister Eckhart put it: "It is a remarkable thing that what flows out remains within. That the word flows out and yet remains within." In these two connected statements, Eckhart discloses mysteries so profound as to be fundamental to both science and theology – namely that inwardness or intensional reality is essentially inexhaustible, no matter how many forms it 'flows out' into. The soul, like the inwardness of the word, is not something to be found within its extensional, physical form, but belongs to an unbounded non-extensional realm of inwardness as such. It belongs to the realm of intensional reality, not to the insideness of extensional bodies.

Understanding the nature of inwardness or intensional reality provides the bridge between Fundamental Science and Fundamental Theology, allowing us to understand Fundamental Science as Fundamental Theology and vice versa. God is no actual thing, nor even every-thing, but an awareness of boundless inner potentiality – of an inexhaustible realm of inwardness that flows out into every actual thing and from which all things flow out. Potentiality has reality only in awareness. Intensional reality, scientifically and theologically, is the reality of potentiality in awareness. Awareness as such is not, in the first place, awareness of any actual thing, but awareness of potentiality. It is only through awareness of potentiality that potentiality can be actualized. Potentiality, possessing its own reality in awareness and knowing itself through awareness is power. God is therefore indeed an all-powerful knowing awareness, one that is ontologically prior to any actual being, but constitutes the inexhaustible inwardness or intensional reality of every being. The power of God is indeed within us all, for it is not a power over beings but a power within all beings. Through our own awareness of potentiality, not specific human potentials but pure potentiality, we are linked directly to God. Through it we know God intimately – not as an actual being, an object of knowledge, but as that fundamental knowing awareness that is the source of all beings. Scientifically speaking, agnosticism is not an option. For that intimate knowing awareness *gnosis* is itself the source of any and every thing and being that can be known. True *gnosis* is not knowledge of or about God. It is God. And it is within us.

Inwardness, as an intensional realm of inexhaustible Potentiality cannot be identified with Being any more than it can with actual beings. It is Non-being. And yet Non-Being also is, possessing an intensional reality that transcends Being as Actuality. Agnosticism in its undecidedness as to whether God 'is' or 'is not', rules out in advance the possibility that God is precisely because God is not, that God is the very bridge of knowing awareness that spans the gulf between Non-Being and Being, Potentiality and Actuality, No-thingness and Thingness. That bridge is actualization, the formative activity which not only gives creative expression to an inexhaustible realm of potentiality, but in doing so actively enlarges and enriches that realm, bestowing it with essential inexhaustibility. Awareness is itself the inward source of this formative activity or 'energy', an activity that has no prior agent in the form of a being, even a divine being, but is essentially autonomous. God in this sense, is no ultimate creator being and no creature or created being, no agent of creation and no mere object of man's creation either. God is aware and autonomous action, action that is its own agent and possesses its own intrinsic awareness. Awareness in turn, is not the property of any pre-given subject, human or divine. It is not the property of God. It is God, understood as a singular gestalt of aware activity that is aware of itself in the smallest of actions and interactions that takes place in the realm of Actuality. An aware activity or active awareness that knows itself in and through each of its expressions, but still abides within itself, for it is an awareness that is never lost in its actualizations. An aware activity that, in Eckhart's words allows potentiality and Non-Being to 'flow out' into Actuality or Being, but 'remains within' even as it flows out.

For, as Parmenides already recognized, Non-Being is inherently paradoxical – it is. It is as that intensional dimension of reality that is the foundation both of Fundamental Theology and Fundamental Science. For Descartes there was only *res extensa* and *res cogitans* – extensional reality of bodies in space and time, and the reality of the knowing subject or 'I'. For Fundamental Science and Fundamental Theology both *res cogitans* and *res extensa*, known objects and knowing subjects are an expression of *res intensa* – of self-actualizing fields of aware potentiality, fields whose ultimate source deserves no other scientific designation than 'God'.

The 13th Fundamental Distinction

The 13th fundamental distinction is between 'awareness' or fundamental cognition and 'consciousness' or secondary cognition. Secondary cognition is consciousness of some actual thing. Primary or fundamental cognition is the direct awareness of potentiality – of potential sense or significance (felt sense), potential sounds and signifiers, potential actions and events, potential phenomena or patterns of events, potential patterns of relatedness and potentials within oneself and others. Fundamental Cognition is our direct inner link with intensional reality – with the inwardness of things and people, events and phenomena. The realm of intensional reality is not poorer or less highly differentiated but richer and more highly differentiated than extensional reality. The richness of potentiality vibrates within actuality, and our awareness of it is a form of "inner vibrational touch" (Seth), a resonant inner contact with every actual thing or person. "The relation that constitutes knowing is one in which we ourselves are related and in which this relation vibrates through our basic comportment." (Heidegger). But this is a knowing understood in the fundamental sense – a knowing awareness of potentiality that precedes and pervades its expression in actuality. So-called 'pre-cognition', from this point of view, is not some bizarre and extra-ordinary parapsychological phenomenon. It is that fundamental cognition or *gnosis* that is the very pre-condition for the emergence of phenomena, an inner vibrational resonance with their pre-existent reality as intensional field-patterns of awareness – resounding with their own fundamental tone.

The Semiotic Confusion of Physical Science

A fundamental semiotic confusion reigns in our entire current understanding of 'science', a confusion between empirical facts on the one hand from their representation in verbal signifiers. The actual starting point of all scientific investigation is not a fact, an observed phenomenon in the natural world, but a pre-given concept of that phenomenon in the form of a human word – a verbal signifier such as 'energy', 'mass', 'light', 'gravity', 'heat', 'attraction', 'repulsion', 'field', 'wave', 'particle' etc.

Every such signifier has a human meaning that transcends its significance as a natural-scientific term. Its fundamental significance cannot be reduced to its significance in denoting a particular natural phenomenon. Instead the deeper significance of the observed physical phenomenon which it refers to can only be understood through the corresponding psychical phenomena that constitutes the human meaning of the signifier.

As a signifier, the term 'warmth' for example, refers both to a measurable, physical phenomenon and to an immeasurable psychical phenomenon. The psychical phenomenon is not the warmth of any natural body, not even the human body, but the inner 'soul warmth' of a human being. We have a direct felt sense of the psychical warmth or coolness, levity or gravity, light or darkness of a human being. Our entire relation to our own being and to other human beings is built on this awareness of human psychical or soul qualities, each of which in turn corresponds to a particular outwardly observed physical phenomenon of nature. The feeling cognition of inergetic warmth – the inner soul warmth we feel within ourselves and others – is no more and no less 'subjective' than our awareness of the physical temperature of our own bodies or those of others. The difference is only that the latter is something measurable and quantifiable.

Marx's vision of a single, Fundamental Science that constitutes both a "human science of nature" and a "natural science of man" can only be realized through studying the human body and brain as a natural phenomenon like any other. It can only come about through an acknowledgement of our specifically human nature – the manner in which specific natural qualities such as light, warmth, gravity etc. are felt within the human being.

Any scientific term in the form of a verbal signifier such as 'warmth' points in two directions simultaneously. It has two basic vectors of signification. On the one hand it has a physical signification – referring to an outward property of natural bodies. On the other hand it has a metaphysical denotation, signifying an inner property of human beings. On the one hand it signifies an observable physical phenomenon. On the other hand it signifies an experienced psychical phenomenon. On the one hand it refers to a measurable physical quantity. On the other hand, to a felt psychical quality. On the one hand it describes a form of energy. On the other hand a form of inner energy or 'inergy'.

It is of fundamental importance to recognize however, that inergetic qualities are not phenomena in the ordinary sense of 'things' that we are consciously aware of but are themselves qualities of awareness. When we feel the inner warmth of another human being for example, what we are feeling is warmth that belongs to their own felt awareness of the world and their own felt relation to other people – a felt awareness and a felt relation that we ourselves feel bathed in and warmed by. Physical phenomena are the expression of energetic relationships between bodies in space time. Similarly, psychical phenomena are the expression of relationships between beings, relationships whose very medium is not energy but inergy – qualitative tones, textures, densities and intensities of awareness itself.

To understand our nature as human beings, it is not enough to simply import concepts borrowed from the natural sciences or scientific technologies and apply them to the human body. The way in which today, practitioners of alternative medicine speak about 'life energies', the human 'energy body' or about healing as a 'balancing' of energies constitutes an importation of the term 'energy' from physical science which relies totally on its physical signification, and completely fails to recognize the fundamental distinction between physical energies on the one hand and their psychical or inergetic counterparts on the other.

The confusion between measurable physical quantities and felt psychical qualities, has a long history. No sooner, for example, had natural science begun to give pride of place to the concepts of magnetism and electricity, than these concepts were applied to the understanding of living organisms and to the human being. Already in the late eighteenth century, Mesmer spoke of 'vital' or 'animal' magnetism', just as, later, others would speak of the body's vital electricity or vital 'energy' in general. The use of new scientific concepts or technologies as models of human nature (for example seeing the brain as 'computer') is a creative process but at the same time a very one-sided approach to the relation of natural phenomena and human nature. For it involves transferring to the human being the physical signification of these concepts in a way that completely ignores their own non-physical or psychical dimension of signification that is central to an understanding of the human being. As a result, the concepts themselves, whilst invariably treated as fundamental to an understanding of human nature, are not themselves transformed into fundamental concepts.

To transform concepts such as 'magnetism', 'electricity', 'warmth', 'light', 'gravity' etc into fundamental concepts means grasping their twofold significance – not just as signifiers of observable physical phenomena but as signifiers of independent psychical phenomena. Their fundamental significance is not grasped by interpreting human nature in terms of their physical signification alone. Applying concepts drawn from the natural sciences to understand our own human nature does indeed lead to a 'natural science of man'. But this must be complemented by a type of research that proceeds in quite the opposite direction and leads towards a 'human science of nature'. This is a science that enriches our understanding of natural-scientific concepts and of the very physical phenomena they refer to through a direct psychical exploration of our own human nature. The method of research that leads to a natural science of man is physical observation of the human being as an expression of natural phenomena. The method of research that leads to a human science of nature is the human being's own inner feeling cognition of nature as the expression of psychical phenomena and psychical reality – of patterned tones and textures, densities and intensities, directions and flows of awareness as such. For within these are to be found the 'archetypes' or inner field-patterns of significance, which find expression not only in the basic signifiers of science but in the physical phenomena they refer to.

Fundamental Research is based on a feeling cognition of the inner psychical counterparts to the physical phenomena signified by basic scientific terms. But it takes as its starting point not just specific scientific terms but the linguistic field of signifiers surrounding them and the fundamental polarities associated with them. Taking 'warmth' as our example, the fundamental polarity is warmth and coldness. The field of signifiers includes phrases connecting psychic warmth and coldness with the heart and with the blood – phrases such as 'warm-hearted', 'cold-hearted', 'hot-blooded' or 'cold-blooded' for example. These in turn have their own physical-scientific significance – for we know that as mammals we are warm-blooded and that we owe this bodily warm-bloodedness to the functioning of our hearts. But warmth and coolness as qualities of the human being rather than the human body are associated with inner closeness and distance, that is to say with a basic polarity of intensional space. Inner warmth and coolness also have a dynamic character – the experience signified by expressions such as 'warming to someone' or 'going cold on someone'. To regard an expression such as 'inner warmth' as merely metaphorical is to assume such experiences are less real than the experience of physical warming up in front of a fire, or cooling down on a wintry day. To seek some scientific or esoteric phenomenon 'behind' such experiences is to assume that they are only real if they can be identified with some physical or quasi-physical force or energy.

For refugees or camp inmates of a Siberian labour camp abandoned to winter cold their bodily survival is dependent as much – if not more – on the inner warmth of their human contact with one another as on the warming effects of their bodies on the air around them. Terms such as 'inner warmth', 'inner light' etc. are not metaphorical signifiers. Nor are they literal signifiers of a physical or quasi-physical energy. Our sense of the inner tone and timbre, warmth and coldness, lightness and heaviness, distance or closeness of another being is something so close to us, so intimate a part of our everyday 'exoteric' experience that science ignores it and theosophy treats it as something esoteric. Ultimately it requires no abstruse theosophical terms such as 'warmth ether' or 'aura' to feel the reality of a person's inner warmth of feeling or the invisible radiance they emanate, for these are simply sensual qualities of feeling awareness belonging to their body of feeling awareness.

Fundamental Research

Physical science takes as its object physical warmth and coldness. Fundamental Research has as its object intensional, inergetic or inner warmth and coldness. It knows this object not through physical observation but through feeling cognition. But this subjective or feeling cognition then in turn facilitates a no less 'objective' observation and description of inwardly cognized psychical phenomena than that which physical science offers for outwardly perceived physical phenomena. But it also brings with it something more – a feeling cognition of the correspondences between different psychical phenomena, which can add to and illuminate the relation between their outer physical counterparts. In our awareness of a human being, for example, felt qualities of darkness, density, heaviness and gravity go together. So also do their polar complements – qualities of 'lightness' in the form of airiness, levity and transparent lucidity or brightness of spirit. For the physical scientist does not begin by taking heed of language. It seems of no consequence that the verbal signifier 'light' has a double sense of luminosity and levity, still less that our felt sense of luminosity and expansion goes together with a felt sense of lightness and levity. As a result, despite the inherently impersonal and collective nature of physical-scientific research it required the extraordinary personal and psychological qualities of an individual scientist – Einstein – to intellectually intuit an inner connection between light and gravity.

Scientific Socialism and Spiritual Science

Whilst Einstein is a name familiar to all as that of a 'genius', the name Rudolf Steiner (1861-1925) has for decades been unmentionable in academic and scientific circles. No greater insult can be afforded to a great thinker than for him to be ignored. The refusal of the academic and scientific world to make even a passing reference to his role in the history of philosophy and science is equivalent to the way in which the figure of Trotsky was erased from all photographs of revolutionary leaders in the Stalin era. It was Rudolf Steiner, who amidst the clash of materialism and spiritualism, science and religion, Western-based philosophies and Eastern-based theosophies that so characterized the culture of the late nineteenth and early twentieth centuries, first clearly formulated the concept of a 'spiritual science' or 'soul science' that could lay the foundations for a new understanding of the material world and physical phenomena. But Steiner went further even than this, being the first to define, develop and apply specific methods of spiritual-scientific research that acknowledged a fundamental distinction between intensional and extensional reality – between soul-space, soul-forces and soul-qualities on the one hand, and physical spaces, forces and quantities on the other. He understood spiritual-scientific research not simply as psychology in the limited, modern sense of this word but as inner psychical research into outer physical phenomena, as objective research into subjective dimensions of reality as such. Like Marx, Steiner envisaged a single 'fundamental' science that would unite the natural science of man with a human science of nature. Steiner called this unity 'Anthroposophy'. Marx called it Scientific Socialism. And just as Marx had linked science with social transformation, and saw socialist thinking as something deeply scientific in character, so also did Steiner see Anthroposophy as something deeply social in character – planting the seeds of social transformation. Marx had seen in humanity's formative relationship to nature the basis of human relations. He understood the essence of human nature itself as sensuous activity embodied both in creative labour and in creative and recreative activities of all sorts. Steiner understood the essence of human nature as spiritual activity manifest in thought but capable also of leading awareness into suprasensible dimensions of fundamental reality.

From a modern scientific perspective, the world outlooks of earlier pre-scientific civilizations appear to have been dominated by a form of 'anthropomorphism' which endowed nature with human attributes and thus saw it as a creation of human-like spirits or gods. Science claims to have rejected 'anthropomorphism' of all sorts – the projection of human features onto natural phenomena. And yet it is anthropomorphic in the deepest sense – treating our own species-specific forms of perception as 'higher' and more fundamentally 'real' than that of all other species. It is also anthropocentric in the extreme – eager to find out whether life exists elsewhere in the universe but prepared to recognize that life only if it takes visible form or manifests to our own human sense organs and mode of perception.

Modern anthropology has almost entirely lost the capacity to understand that in the very earliest eras of human development, people did not psychologically 'project' onto nature their own human psychical qualities – 'romantically' projecting human feelings such as anger or gentleness, for example, onto natural phenomena such as storms or streams. The truth is quite the contrary. They experienced their own human emotions and other psychical qualities as natural forces. Conversely, however, they were still capable of allowing their own awareness to flow into natural phenomena – to enter into and flow with a stream or storm, wind or waterfall. They possessed an innate feeling cognition of nature that required no scientific language to represent it. Instead nature itself spoke for them. They experienced the wind and storm without as the larger voice of a wind and storm within, outer nature as an expression of their own inner nature, and the outer universe as the manifestation of an inner universe that was the source of both nature and man. In particular they experienced their own outward motions of awareness – their 'e-motions' – as a medium through which awareness could flow beyond the boundaries of the body and into natural phenomena themselves – into light and air, fire and water, plant and rock. Steiner's anthroposophical world outlook is not an outdated form of 'anthropomorphism' characteristic of earlier cultures and societies but a scientific insight that there is an innate resonance and isomorphism between the inner being of nature and the inner nature of the human being. That the inner universe is a universe of beings and not of bodies in extensional space.

Today there is no longer any scientific understanding of the way in which personal feeling, if followed, can lead into pre-personal and trans-personal dimensions of feeling cognition that open our awareness to pre- and trans-physical dimensions of reality itself. Therefore nothing is more foreign to science than the idea that our life of feeling should have any role to play in the quest for knowledge of the universe – for it is identified with the realm of the purely personal. The scientist must thus be someone who sacrifices individuality and personal feeling to a coldly impersonal and dryly intellectual rationality and to universal standards of clinical objectivity. The results are all too clear. In medicine and psychiatry the individual patient is reduced to a statistical case of some generic disease following its own genetic laws. In what passes as 'scientific' the feeling life of the human being is reduced to an object of psychometric testing and quantification. In psychoanalysis it is reduced to an archaic underworld of unconscious libidinal passions, desires and impulses. The idea of an impersonal universe governed by universal laws is reflected socially in a fetishism of impersonal laws of the market and the imposition of an impersonal rule of law necessary to deal with the unwanted disruptions of the social order stemming from personally motivated crimes and misdemeanours. This is a society, in which, as Marx so clearly pointed out, impersonal relations between things – technologies and impersonal market 'forces' – dominate relations between people. It is also an inherently unstable society in which the protection of impersonal economic, political and scientific laws is required to contain and subjugate the human subject – the source of unruly personal passions and impulses.

At the heart of the scientific and social rejection of feeling cognition and its substitution by manipulation and control, is a fundamental confusion between feeling as such and personal feelings – the latter being seen as 'things in themselves'. The word 'feeling' is first and foremost a verb and not a singular or plural noun. To feel something or someone is to be in direct contact with them. In holding a mug of coffee the way I feel the mug itself in my hand is not reducible to a passive and purely 'personal' feeling in my hand as it holds the mug. But the way the mug itself feels does indeed depend on the way I hold and feel it. Intellectual cognition is a representation of relationships between things or between people. Feeling cognition has to do with our own active relation to things and to people – the depth and breadth of our direct feeling contact with them.

Someone deprived or incapable of warmth of feeling cannot at first feel this warmth in another. Steiner's spiritual-scientific world view and research methodology was founded on the recognition that the active cultivation of feeling cognition was a necessary precondition for its scientific application. His genius lay in recognizing that the very subjective faculties necessary for spiritual scientific research must bear the same fundamental nature as their objects. Thus a coldly impersonal science cannot, by his or her very nature, obtain a feeling cognition of inner warmth. An individual scientist dominated by a cool, clinical approach to a particular object of investigation, cannot transform the heat of passionate personal interest into the warmth of soul necessary for genuine feeling contact and cognition with that object. Nor can they transform this feeling contact and cognition, borne of inner warmth, into a lucidity of thought that still bears, like sunlight, this warmth within it. It is the self-imposed separation between the clear light of intellectual cognition on the one hand and inner warmth of feeling cognition on the other – a separation that has its source in the very person of the scientist – that, paradoxically, lends science its impersonal character. The impersonal character of the scientific intellect disguises in other words its own deeply personal source in the split personality of the scientists. But a psychologically split personality cannot be the source of a unified science. Hence the emphasis that Rudolf Steiner laid on the unification of three quite distinct but nevertheless inseparable modes of cognition – thinking, feeling and willing, and on the cognitive processes necessary to link these modes.

The world today is characterized by an uncentred polarization of the repressed fire of authentic spiritual will impulses burning in Islam and the cold light of scientific technologies employed by the West as an instrument of manipulation and control, commercial gain and military-economic power. Steiner identified the uncentred or heartless polarization of unfeeling intellect and unfeeling will as the essence of 'evil'. It cannot be overcome by a sentimental Christianity of the heart that does not recognize the scientific and social significance of feeling cognition. For evil's ultimate source is not the evil intents of individuals but their spiritual-scientific ignorance. That is why Steiner saw the development of spiritual science as something that should be the concern of everyone and not just scientists or those with an interest in science. For its essence lay in enabling the individual to transform spiritual values and will-impulses, through the warmth of soul embodied in the heart, into clear-headed thinking. And vice versa, to transform otherwise cold and impersonal scientific concepts, through a warmth-filled feeling cognition, into spiritual deeds. Spiritual science meant the acknowledgement of feeling cognition as the centre of two fundamental cognitive processes that together constituted a cognitive cycle. One process leading from the will to the warmth-filled word or signifier and the other leading from the word or signifier to warmth-filled worldly deeds.

The psycho-social issues that Steiner tackled, and the cognitive processes he saw as necessary to tackle them, are echoed in all manner of ways in the theory and practice of psychotherapy. Within the culture of psychotherapy, however, individual therapists tend to see themselves first and foremost as qualified professional practitioners – there simply to apply whatever methods and models were handed down to them in their professional training. Whilst working at the cutting edge of psychology, they do not see themselves as psychologists – that is to say, as scientists of the psyche. A true psychologist is first and foremost a scientist – someone whose every experience of themselves and of other human beings can affect and alter their general understanding of the human soul or psyche. To be a psychotherapist without being a psychologist – without being a genuine scientist in this sense – is indeed a contradiction in terms. It means to reduce the psychotherapist to the status of a cult follower – the 'professional' practitioner of an institutional culture built around its own fixed models, methods and metaphors. Since Freud, psychotherapy, despite its cult nature, has become an accepted part of social culture. Rudolf Steiner on the other hand, is seen only as the founder of a theosophical cult (Anthroposophy) and of a specific mode of education (Waldorf education). This is indeed a paradox, given that what Steiner himself taught was what it meant for an individual – any individual – to become a genuine scientist of the soul, and in this sense a true psychologist rather than a cultist.

Traditionally, theosophy has had in common with science and theology a tendency towards a literalistic interpretation of its own language, one which implies a one-to-one relation between words and pre-given things, between verbal signifiers and the phenomena they signify. Rudolf Steiner was not immune to this tendency, for in order to simplify the results of his own spiritual-scientific research he spoke constantly of the relation between the four basic 'members' that made up the human being: the 'physical body', 'etheric body', 'astral body' and inner being or 'ego'. In doing so he made use of an existing theosophical terminology but was also forced to deviate somewhat from the basic principles of his own anthroposophical world outlook. To describe the characteristics of three distinct bodies, without, as Heidegger did, questioning and refining the fundamental concept of bodyhood as such, is to fall into what I call 'adjectival thinking'. It contrasts with Steiner's own call for a dynamic thinking of the sort that Heidegger made use of, one which found expression in his understanding of bodyhood not as a thing but as a dynamic process of bodying. Steiner himself was well aware of the tendency amongst theosophists to objectify their own concepts with the result that spiritual or intensional reality was represented in material or extensional terms. Above all, he emphasized the social and scientific significance of establishing a personal relation to the otherwise impersonal concepts of science, one through which their intensional meaning and spiritual significance could shine through. Like Heidegger in other words, he recognized that a deepened inner relationship to the world went hand in hand with a deepened inner relationship to the word. Following Heraclitus, Heidegger understood this new relationship as a listening relationship to language, one through which words themselves could speak to us in a new way, and reveal inner relationships and resonances that transcended their formal reference.

Fundamental Science is a modern field-dynamic and field-phenomenological articulation of the principles and practice of spiritual science as Rudolf Steiner understood it. The difference between Fundamental Science and Anthroposophy lies only in the greater emphasis it places on the semiotic dimension of spiritual knowledge, and in particular on the researcher's own relationship to language. I must emphasize however, that Rudolf Steiner was personally well aware that there were fundamental questions to be asked surrounding the relation of language and sign systems to inner knowing and intensional reality, and himself suggested answers to these questions which are fully in resonance with those offered by Fundamental Science.

The research methodology of Fundamental Science is founded on a demystification and demythologization of the distinction between extensional and intensional reality, physical signifiers and their psychical sense. Physical phenomena such as warmth, light, gravity and electromagnetism are understood as physical signifiers of psychical phenomena – material metaphors of inner warmth, light, gravity etc. The latter in turn are understood as different inergetic dimensions of feeling tone. For just as physical tones can have qualities of lightness and darkness, warmth and coolness, so can feeling tones. Feeling tone is not a 'thing in itself' but the very medium of interrelatedness between beings. Fields of awareness are fields or planes of interrelatedness – not between bodies in extensional space but between beings. Feeling tones are the patterned tones and intensities, directions and flows of awareness that constitute such fields of interrelatedness. Research into psychical phenomena such as sensed inner warmth, light and gravity therefore has itself a necessarily relational character.

The principle method of Fundamental Research is the use of the field between two or more people to enter and explore the intensional fields or planes of awareness that unite them as beings, and in this way enter the different 'inergetic' tonalities and textures that constitute these fields or planes of reality. But Fundamental Science, like Spiritual Science, has another dimension too. That is the understanding of physical phenomena and energetic fields as phenomenal signifiers or material metaphors of psychical phenomena and fields of awareness.

Both verbal signifiers (the word 'warmth') and physical signifiers (the physical phenomenon of warmth) have a sensed psychical significance. The sensed significance of physical signifiers – for example physical sensations of warmth and coolness – lie in the way they evoke or reflect, correspond or contrast with inner psychical warmth or coolness that the individual feels, or does not feel within themselves. The psychical significance of verbal signifiers is sensed through their inner resonance. Not just the words 'warmth' and 'coolness' but any words or chain of signifiers may carry a specific resonance that induces a felt sense of warmth or coolness. This sense may find phenomenal expression in both physical and verbal signifiers, both waking and dream experiences Thus a warm feeling of security and well-being or a cold feeling of abandonment may be signified through a dream experience of a warm tropical beach or a cold and empty building. It may also be signified through an actual waking experience of such a beach or building, or simply through physical sensations of warmth and coldness as such. It can convey itself verbally and physically through the warmth or coldness of a person's words and their physical tone of voice. Finally, in dreams, physical signifiers may symbolize verbal signifiers – as in the metaphorical word–image or 'rebus'. Thus a person may dream of a kettle going 'off the boil', using this physical signifier to signify an inner sense belonging to this verbal expression – a felt loss of passion or heat of involvement with something or someone.

What is the essence of 'warmth'? Is it an experience, a form of physical energy or, as it was originally understood in both East and West, a primordial element? The idea that the essence of warmth is 'inner warmth' – the intensional warmth of a being rather than of an extensional body is reflected in Rudolf Steiner's direct association of the element of warmth with the inner human being. Our warmth-being, for Steiner, is our inner being. Any being is essentially a warmth entity. But since warmth, light, gravity and sound are all, from the point of view of Fundamental Science, different tones and textures of awareness, we can also understand Rudolf Steiner's reference to a warmth 'ether' as a basic medium of interrelationship between beings. Fundamental or essential warmth, as psychical or inner warmth is something we can gain a feeling cognition of through our own warmth-being – our nature as warmth entities inhabiting a non-physical warmth medium or ether. Similarly fundamental or essential light is something we can gain a feeling cognition of only through our own nature as beings composed of an inner light and inhabiting the open space of that light – that light of *awareness* within which alone anything first becomes visible.

For Fundamental Science, it is awareness, not extensional space, that constitutes the basic medium of existence. Awareness however, is imbued with its own felt tonalities, and fields of awareness consist essentially of patterned flows and figurations of feeling tone. There is no more ludicrous an expression of the physical-scientific world outlook than the idea that the enjoyment of music is a result of subjective feelings induced by mechanically generated tones – vibrations of air molecules. All previous attempts to create a psychology or phenomenology of musical experiences have floundered through their failure to comprehend a single fundamental principle: namely that feeling tone is something more primordial than either psychical feelings or physical tones. Feeling tone is the primordial source of musical composition and that which is embodied in music making. It is also the very medium of musical resonance. The expression of feeling tone in the form of mechanically or vocally generated tones not only gives it audible vibratory form but sets up an amplificatory resonance between patterns of musical tone and the patterns of feeling tone that are their source. The outer tones that are produced in music making serve only as the medium of a direct inner resonance with the inner music of feeling tones that is its source. These are not just matters of interest only to musicians, musicologists or music lovers. The inner universe is fundamentally a musical universe. Inner field-patterns of awareness, as patterned tonalities of awareness – patterns of feeling tone – are the music of the inner universe. Conversely, the experience of music as we know it is itself the most direct expression of the nature of the inner universe – the nature of intensional reality.

Extensional bodies, including the molecules of air set in motion by music, are governed by laws of movement in extensional space – by momentum and inertia. Movement in intensional space has a quite different character. It is not governed by relationships of momentum and inertia but by the relationships of flow and form. Movement or *kinesis* in the original Greek sense is metamorphosis. Intensional movement is not change of place but change of form – transformation or metamorphosis. The flow of music is a metamorphic or morphodynamic flow – consisting of changing forms or figurations of feeling tone. From out of a sea of flowing tonalities of awareness emerge whole landscapes of feeling tone, which may be perceived, visualized or dreamt as extensional, natural landscapes. Patterned flows and figurations of feeling tone are the musical infrastructure of the inner universe. Feeling tone as such is the basic 'ether' or medium of awareness, of which these flows and figurations are composed. Just as physical sounds have their own felt qualities of warmth and coolness, brightness and darkness, lightness and heaviness, softness and hardness, angularity or roundedness, harmony and dissonance, colour and shape, so are all actual qualities of the physical world as we experience it expressions of inner or intensional sound – shaped expressions of feeling tone.

Form is stabilized flow, in resonance with its own inner field-patterns of tonality. Any phenomenal form whatsoever is essentially a flowform. But flow itself, as in music, is essentially a fluid transformation or formflow – the dynamic metamorphosis of one formed expression of feeling tone into another. When we are in resonance with music, our own awareness flows into its form. That is to say, our own felt tonalities of awareness are not only echoed in the musical tones but take on the flowing forms of the music itself.

Rudolf Steiner defined three basic modes of spiritual or intensional cognition which he named Imagination, Inspiration and Intuition respectively. His understanding of these three key terms is related to, but also transcends their conventional signification. It can most easily be understood through reference to musical cognition as the morphodynamics of feeling tone. Steiner's understanding of 'Imaginative' cognition for example can be compared to the images evoked by music – not actual music but an inner music that is felt without actually being heard. What he called 'Inspiration' corresponds to what happens when we pass from the realm of imagination, dreams and memories to a direct awareness of the feeling tones to which they give form. 'Intuition', on the other hand, is comparable to what happens when our awareness flows into and along with those tones, itself taking on different forms. The latter may in turn be experienced once again through Imagination, but a type of Imagination in which our own awareness is within the 'images' we behold. Intuition is essentially movement in intensional space, understood as a movement through feeling tone. In listening to a piece of music there is a sense in which the music flows through us. But there is another deeper sense in which our awareness flows not only 'with' the music but through the tones of feeling echoed in it. In life too, there is an essential difference between experiencing different moods or tones of feeling as they pass through us, and consciously passing through one mood or tone of feeling to another. This in turn depends on our capacity to experience feeling tones, not as emotions or sensations that we find within us, but as basic moods that we find ourselves within – permeating and colouring our entire field of awareness, inner and outer.

Essence, Elements, Ethers and Energies

What is the essence of 'warmth'? Is it an experience, a form of physical energy or, as it was originally understood in both East and West, a primordial element? The idea that the essence of warmth is 'inner warmth' – the intensional warmth of a being rather than of an extensional body is reflected in Rudolf Steiner's direct association of the element of warmth with the inner human being. Our warmth-being, for Steiner, is our inner being. Any being is essentially a warmth entity. But since warmth, light, gravity and sound are all, from the point of view of Fundamental Science, different tones and textures of awareness, we can also understand Rudolf Steiner's reference to a warmth 'ether' as a basic medium of interrelationship between beings. Fundamental or essential warmth, as psychical or inner warmth is something we can gain a feeling cognition of through our own warmth-being – our nature as warmth entities inhabiting a non-physical warmth medium or ether. Similarly fundamental or essential light is something we can gain a feeling cognition of only through our own nature as beings composed of inner light and inhabiting a light ether. Put simply, feeling warm is a sensory experience, but warmth of feeling is something quite different – a psychical quality or *quale* rather than a physical one. Such psychical qualia are not merely sensations or sensory qualities we are aware 'of' but innately sensuous qualities *of* awareness – such as the light of awareness that may shine out through our eyes, or the warmth of feeling awareness that may emanate from our bodies or resound in the tone of our voice.

Our felt sensation of particular physical phenomena and the felt resonance of the verbal signifiers attached to them can lead us to a direct felt sense of their psychical significance. Through it we experience both physical and verbal signifiers as expressions of primordial psychical qualities. But sensed significance or felt sense is a gateway to direct feeling cognition of these qualities as dimensions of our own being – dimensions of the felt self. The latter is a field self composed essentially of feeling tones. It is the sensed qualities and textures of these feeling tones that bestow them with a quasi-sensual character – an 'energetic', 'etheric' or 'elemental' character.

Through sensual awareness of light, darkness and colour we also sense their inner psychical counterparts. The felt meaning of light, darkness and colour as we experience them through a dull, grey sky is in resonance with verbal expressions such as 'feeling dull'. But we could not feel a relation between a sensed dullness of mood and the dull greyness of the sky above us were they not in some way *both* expressions of the dulling of a different and more fundamental type of light – the light of awareness. Similarly, if we feel in a bright or radiant mood it is this light which shines within us and, at the same time, lends also a greater radiance to our awareness the world around us.

The path of Fundamental Research is a path that leads from personal experiences of nature and personally felt resonances of language to something far deeper – to a direct feeling cognition of an 'etheric' or 'elemental' world manifest in our field states and felt tonalities of awareness. For lighter or darker, brighter or duller moods are also the expression of lighter or darker, brighter or duller basic *tones* of or awareness. It is these tonalities of awareness which lighten, darken and generally 'set the tone' of our moods as well as colouring our experience of the world.

If somebody were to detect a dull or dark mood or 'feeling tone' within us and ask us to 'lighten up' or 'brighten up', this is something we will feel disinclined to do - feeling it as a demand to put on a superficial and inauthentic mask of brightness and cheer. And yet we each do possess the capacity to lighten or brighten up in an authentic way – to use our own intent to alter the felt tone of our being, in a similar way that we do when we brighten or lighten our tone of voice. We do so using our own human organism – that instrument or *organon* with which we modulate not just our vocal tones but the silent tones of feelings they resound with – the inner *music* of the soul.

What Rudolf Steiner calls our 'warmth organism', 'fluid organism', and 'air organism' are not separate bodies or 'members' of our own organism so much as elemental ways of sensing our own organism as a body of awareness – for example sensing it as fiery or airy, fluid and watery, or dense, solid and 'earthy'. For just as vocal or musical tones can have sensed and sensual qualities of lightness or darkness, sharpness or dullness, warmth or coolness, hardness of softness, earthiness or fieriness etc. – so can and do all the silent *tonalities of awareness* whose innately sensual qualities or psychical 'qualia' they express or resound with.

Fundamental Science and Music

For Fundamental Science, it is awareness, not extensional space that constitutes the basic medium of existence. Awareness however, is imbued with its own felt tonalities, and fields of awareness consisting essentially of patterned flows and figurations of feeling tone. There is no more ludicrous an expression of the physical-scientific world outlook than the idea that the enjoyment of music is a result of subjective feelings induced by mechanically generated tones – vibrations of air molecules. All previous attempts to create a psychology or phenomenology of musical experiences have floundered through their failure to comprehend a single fundamental principle: namely that feeling tone is something more primordial than either psychical feelings or physical tones. Feeling tone is the primordial source of musical composition and that which is embodied in music making. It is also the very medium of musical resonance. The expression of feeling tone in the form of mechanically or vocally generated tones not only gives it audible vibratory form but sets up an amplificatory resonance between patterns of musical tone and the patterns of feeling tone that are their source. The outer tones that are produced in music making serve only as the medium of a direct inner resonance with the inner music of feeling tones that is its source. These are not just matters of interest only to musicians, musicologists or music lovers. The inner universe is fundamentally a musical universe. Inner field-patterns of awareness, as patterned tonalities of awareness – patterns of feeling tone – are the music of the inner universe. Conversely, the experience of music as we know it is itself the most direct expression of the nature of the inner universe – the nature of intensional reality.

Extensional bodies, including the molecules of air set in motion by music, are governed by laws of movement in extensional space – by momentum and inertia. Movement in intensional space has a quite different character. It is not governed by relationships of momentum and inertia but by the relationships of flow and form. Movement or *kinesis* in the original Greek sense is metamorphosis. Intensional movement is not change of place but change of form – transformation or metamorphosis. The flow of music is a metamorphic or morphodynamic flow – consisting of changing forms or figurations of feeling tone. From out of a sea of flowing tonalities of awareness emerge whole landscapes of feeling tone, which may be perceived, visualized or dreamt as extensional, natural landscapes. Patterned flows and figurations of feeling tone are the musical infrastructure of the inner universe. Feeling tone as such is the basic 'ether' or medium of awareness, of which these flows and figurations are composed. Just as physical sounds have their own felt qualities of warmth and coolness, brightness and darkness, lightness and heaviness, softness and hardness, angularity or roundedness, harmony and dissonance, colour and shape, so are all actual qualities of the physical world as we experience it expressions of inner or intensional sound – shaped expressions of feeling tone. Form is stabilized flow, in resonance with its own inner field-patterns of tonality. Any phenomenal form is essentially a flowform. But flow itself, as in music, is essentially a fluid transformation or formflow – the metamorphosis of one formed expression of feeling tone into another. When we are in inner deep resonance with a piece of music, our tones and chords of feeling *flow into* it's the form of its musical tones and chords to be taken up, amplified and transformed by them. Music is 'morphic resonance' in its essence – a relation of resonance between *form and feeling tone.*

Intelligent Entities

There is a childish naivety to the belief that only human beings or the 'higher' animals possess intelligence, and in the scientific quest to discover whether there are other planets in the universe where intelligent life exists. The scientist of today carries with them the conviction that, unless there is physical proof to the contrary, human beings are uniquely endowed with an intelligence that enables them to understand the laws of the cosmos. There is no sense whatsoever that human intelligence is but the limited human expression of a cosmic intelligence behind those laws – an expression moreover, that is limited by its own largely unquestioned laws, its own intellectual logic and languages. It is as if a musicologist were to come across a score of a great symphony and to analyse its structure and 'laws' without for a moment considering that that score might be the work of another intelligent being – not to mention an intelligence greater than their own. The modern scientist refuses to even consider that the known universe is the universe of our current human awareness – and that there might be beings whose awareness of the cosmos is deeper and broader than our own, and for whom what appear to our limited human intelligence as fixed laws have no more universal validity than the grammatical rules of a particular language. Rudolf Steiner's belief in the existence of trans-human or 'spiritual' beings with a higher consciousness and intelligence than our own is treated by science as something no less heretical and offensive to human dignity than the Copernican proposal that the earth might not, after all, be the centre of the physical cosmos. Still more challenging is the idea that such 'extra-terrestrial' beings are not to be sought on other planets in space but inhabit an intensional space composed of intensional fields or planes of awareness.

The standpoint of the conventional scientist towards these larger, trans-human intelligences can be compared to that of a student of music composition who fails to understand that the very person of the composer is itself but a personification of that higher consciousness or intelligence which constitutes their own larger musical soul.

All true cognition of intensional or 'spiritual' reality can only derive from their feeling cognition of its inner music. Yet this feeling cognition reaches only as far as the individual's capacity to (a) feel their own personhood as one expression of this music (b) feel their own intellectual cognitions as single strands or patterns within the fabric of a larger soul or consciousness inhabiting a larger field and higher plane of awareness.

By immersing ourselves in the music of a great composer or the language of a great thinker, we gain more than just an understanding of the personality of this thinker or composer. We gain an understanding of their personality essence as the 'signature' of a higher consciousness and intelligence in its own right – the 'great soul' or *Mahatma* of the composer.

Deep listening – into and beyond inner silence – is what first opens up the world of *feeling tone* and its multiple forms, patterns and qualities, allowing a musical *interweaving* of personal human feelings with their *source* in trans-personal and trans-human tones, qualities and patterns of feeling awareness.

It is in this way that, in Rudolf Steiner's terms, we come to 'cross the threshold' to the 'spiritual world' – learning to attune to and resonate with the trans-personal tones and textures of feeling awareness that constitute the musical 'signature' of higher, trans-human beings and consciousnesses.

Appendix - 'Dark Matter' and the Collapse of Physics

As explained in my book *The Science Delusion*, the so-called 'scientific revolution', far from transcending speculation and placing knowledge on a genuinely 'empirical' basis did quite the opposite – treating its own abstract mental and mathematical *concepts* as more 'real' than the tangibly experienced empirical phenomena they were supposed to explain.

Thus, as Husserl argued in his ground-breaking work on *The Crisis in the European Sciences*, the idea that natural science is 'materialist' or 'empirical' is a myth. Instead what is taken as 'natural' or 'physical' science substitutes *".. a world of idealities for the only real world, the one that is actually given through perception, that is ever experienced and experienceable – our everyday lifeworld".*

Husserl here follows in the footsteps of Bishop Berkeley, who first saw through the myth that science offers us a 'harder', more 'solid' account than religion of our actual sensory experience of phenomena. Which is why Heidegger insisted that: *"Phenomenology is more of a science than natural science is."* For 'phenomenology' is that approach to science which explores our direct subjective experience of phenomena whilst at the same 'bracketing' all our mental concepts of it – and recognising them as just that – as mental concepts. This applies above all to the very concept of 'matter' itself – whether visible matter or 'dark matter'. For whilst we subjectively experience the sensory qualities of so-called 'material' phenomena – qualities such as heaviness or lightness, hardness and softness, shape and texture, colour and sound – we never experience or perceive 'matter' as such – as an object or substance. Instead as Samuel Avery argues:

"We experience visual and tactile perceptions that suggest a material substance existing independently, but its acceptance as ultimately real is an act of faith."

Samuel Avery *The Dimensional Structure of Consciousnes*

The still-enduring myth that physical science is in any way 'materialistic' is rooted in the myth of some sort of physical or material 'substance'. In contrast, from an everyday 'phenomenological' or experiential perspective – a truly 'empirical' perspective:

"The concept of material substance ... is derived from potential perceptions in each sensory realm." (Avery).

In particular, we come to think of objects as 'material' only because we do not just perceive them visually but also as something we can potentially touch, hold and pick up and generally come to sense in a tactile way.

What we think of as 'matter' is real therefore only in the root sense of the word – being the 'mother' or 'matrix' [mater] of all things – a 'womb' of potential dimensions of sensory experiencing – of which the tactile dimension is crucial for our identification of things as 'material' in the conventional sense.

The understanding of matter as something inherently connected not with the purely visible or measurable but with an invisible realm or womb of potentiality was long accepted by philosophers and theologians alike. Thus Aristotle defined matter (Greek *hyle*) as potentiality and its form (*morphe*) as actuality. Similarly, Thomas Aquinas understand 'primary matter' (*Prima Materia*) as nothing actual or 'substantial' but as *pure potentiality* – a type of formless and 'passive potentiality' inseparable from God as 'active potentiality'.

This ancient understanding of the essence of 'matter' as such – as something belonging to the realm of the potential rather than the actual – and thus innately invisible and immeasurable – is now echoed in the defining characteristics of what modern physics calls 'dark matter' – a concept which treats it as a mere mysterious and invisible sub-species of measurable or visible matter.

What we must first of all recognise in current scientific approaches to 'dark matter' is that they are essentially nothing but a new 'matterphysical' accounts of anomalies in current physics – invisible sources of gravity. The paradox is that this new 'matterphysical' construct of 'dark' matter simply re-interprets what was previously understood as the metaphysical essence of matter as such.

Whence the need for such a new 'matterphysical' construct? The reason lies in a need to maintain the overall framework of constructs that currently constitutes physics as such in the face of data which threaten this framework and its chief function – that of maintaining a global technological culture in which there are no longer fundamental *philosophical* questions requiring deeper answers – but only 'problems' in need of technical 'solutions' through new technological or military apparatus.

It was Martin Heidegger who first coined the term *das Gestell* (translated as 'the Frame' or 'Frame-up', 'the Enframing' or 'Con-struct') to name and articulate the hidden essence of technological science and of our technological culture. The ordinary German meaning of *Gestell* is some sort of structure, set up or apparatus. By 'technology' however,

Heidegger did not mean actual technological constructions, apparatus, experimental set ups, instruments or gadgets of the sort that have become so much a part of our culture. Nor did he see technology merely as the 'application' of science to the creation of different technologies and their products. Instead he saw technology as the hidden essence of science itself.

The German word *Gestell* derives from the verb stellen – to set, set up or set upon. One of the forms of this verb is *vorstellen* – meaning to 'represent', 'set before' or 'set in front'. *Das Gestell* can thus also be translated as a 'set up' or 'frame up' that seeks to represent something. What is 'framed' by 'The Frame' is like a painting, set before us as a structured representation of something – in the same way the theoretical constructs of modern science are set before us as representations (*Vorstellungen*) of fundamental reality. What such scientific constructs conceal however, is the way in which they themselves are what first frame, define or construct the very idea of whatever it is that science then claims to represent or set before us as some experimentally proven or empirical 'fact'. As Heidegger writes:

"Modern science's way of representing [reality] pursues and entraps nature as a calculable coherence of forces. Modern physics is not experimental physics because it applies [technical] apparatus to the questioning of nature. Rather the reverse is true. Because physics … already as pure theory, sets nature up to exhibit itself as a coherence of forces calculable in advance, it therefore orders its experiments precisely for the purpose of asking [only] whether and how nature reports itself when set up in this way."

Martin Heidegger *The Question Concerning Technology*

Thus it is simply not the case that there is first of all something that exists 'out there' – an 'electron' for example – something which science then happens to have a ready-to-hand term for and ready ways of finding evidence of. Instead the very term 'electron' is a representational construct forming part of the overall framework of physics - just as the understanding that physics offers us as to what 'an electron' *is* is nothing pre-determined by nature, but is instead defined and 'enframed' by that framework.

Were billions to be spent constructing the most sophisticated and expensive technological apparatus, instruments and installations to detect ghosts, most people would consider this an outrageous waste of money. Yet right now there are vastly elaborate technical installations designed to detect what are in effect, no more than ghostly 'particles' invented by scientists to prevent the entire framework of physics from falling apart at the seams.

A current and important example of such constructs is 'dark matter' – an as-yet wholly unexplained source of gravity believed to account for 90 to 99% of the physical universe – yet one needed to 'explain' what it is that stops galaxies from literally flying apart. Together with the concept of 'dark matter' goes the concept of a 'dark energy', supposed to make up 74% of the mass-energy of the universe and uniformly present throughout space. Attempts to identify the nature of dark matter however, postulate in advance its particulate nature. Hence the use of such a massive set up of 'apparatus' (another meaning of *Gestell*) as the CERN Large Hadron Collider to 'discover' the type of particles that make it up

What however, would such a discovery bring – except the confirmation of a postulate already set up – one which prevents nature itself from revealing itself in any other ways besides those already set up in advance by the limited framework of questions on the basis of which it is experimentally interrogated and challenged to 'answer' for itself?

The 'discovery' of a 'particle' that could explain the nature of 'dark matter' would not 'prove' anything except the particular way its nature was pre-conceived in setting up a mode of experimentation within one or more of the models that form part of the current framework of physics. This does nothing to prove that this framework is an accurate 'representation' of reality. For as Heidegger recognised, physics as physics – as a framework of constructs or way of 'enframing' our picture of reality – *is not itself the object of any possible physical experiment.*

No apparatus or modes of experimental measurement by which a 'discovery' of the nature of 'dark matter' could come about could prove anything except the capability of specific experimental apparatus and modes of measurement to limit any possible 'results' in terms of constructs that already 'fit the frame' of current 'physics' and its theoretical models – which delimit and pre-conceive in advance what it is that can be 'discovered'. Such a revolutionary 'discovery' then – for example in the form of a 'particle' such as the Higgs boson – far from being a profound 'experimental breakthrough', would merely permit the addition of one more *theoretically* conceived 'particle' to the current framework of theoretical constructs that constitute physics and predefine every possible experimental 'result' in advance. Nevertheless we see in the urgency to find proof of a new 'particle' the need to prevent the entire theoretically constructed framework of physics from 'flying apart' in precisely the way that 'dark matter' itself is supposed to prevent galaxies from doing! So it comes as no surprise to read that:

"Former Harvard research scholar, professor Shahriar Afshar said that failure to find the particle would bring current scientific theory tumbling down like a house of cards with nothing to replace it." Richard Alleyne

According to Afshar himself:

"There will be an all-out war among physicists. It will be a nightmarish situation that will put physics back into the wilderness."

"We need to start having discussions about what are the alternatives. Because if the LHC [Large Hadron Collider] fails, then the Standard Model fails. If the Standard Model fails we have nothing left."

Amidst all the talk about exciting new 'discoveries' on the horizon relating to 'dark matter', there is not even mention of Einstein's viewpoint that 'particles' as such are not 'hard facts' – and that the very concept of the particle has ceased to serve any purpose:

"Since the theory of general relativity implies the representation of physical reality by a continuous field, the concept of particles or material points cannot play a fundamental part..." and that "... it seems to me certain that we must give up the idea of complete localization of the particle in a theoretical model."

As Heidegger recognised, what today goes by the name of 'science', though it is derived from philosophy, has now effectively replaced even the most elementary forms of philosophical questioning and thinking, the latter now being seen as 'scientifically' outdated.

The problem is that in place of the type of analytic, questioning and critical thinking that once characterised 'philosophy' we now have a type of 'research' whose only possible 'results' are of a sort already pre-defined in advance by the framework of particular 'theories' or 'models'. The theories of science in other words, are judged only according to types of scientific 'evidence' of a sort already framed in advance by those theories and their constructs. Such 'evidence' has no more 'validity' than a box-ticking questionnaire that restricts the one interrogated to choosing from a pre-determined set of answers in response to a pre-determined set of questions – and that according to the pre-determined terms in which the questions themselves are framed. It is in this way that 'scientific' theories and models, together with their supposed 'research evidence' effectively replace, block and ultimately substitute for reflective thinking and questioning – closing off any space for a thinking 'outside the frame'. This is a thinking bound neither to the current mental constructs of science nor to questions framed solely in terms of those constructs, but a thinking that is instead capable of questioning those very constructs and the larger framework of accepted constructs in which they are designed to fit and thereby reinforce.

As Heidegger pointed out in his essay on 'Science and Reflection', it has long since been totally forgotten that the Greek word *eidos* – from which the word 'idea' derives – originally meant an outwardly perceived face or 'aspect' of some thing – and no mere mental 'idea' of it. Similarly, the Greek word *theoria* meant beholding and attending closely to the faces and aspects that things *present* to us in immediate awareness – the very opposite, in other words, of any form of *re-presentation* of things in the form of theoretical ideas or concepts. And Einstein himself – though he became a veritable icon of 'the scientist' – was only too aware that theoretical physics was not so much shaped by evidence *beheld* in immediate awareness as by theoretical concepts *held* in the minds of 'professional scientists' with little or no awareness of the historical and philosophical background of their concepts and theories – and little or no "philosophical insight" into them.

"So many people today – and even professional scientists – seem to me like somebody who has seen thousands of trees but has never seen a forest. A knowledge of the historic and philosophical background gives that kind of independence from prejudices of his generation from which most scientists are suffering. This independence created by philosophical insight is – in my opinion – the mark of distinction between a mere artisan or specialist and a real seeker after truth."

Albert Einstein to Robert A. Thornton, 7 December 1944

"Concepts that have proven useful in ordering things easily achieve such an authority over us that we forget their earthly origins and accept them as unalterable givens. Thus they come to be stamped as 'necessities of thought', 'a priori givens', etc. The path of scientific advance is often made impassable for a long time through such errors. For that reason, it is by no means an idle game if we become practiced in analyzing the long common place concepts and exhibiting those circumstances upon which their justification and usefulness depend, how they have grown up, individually, out of the givens of experience. By this means, their all-too-great authority will be broken."

Albert Einstein 'Ernst Mach'. *Physikalische Zeitschrift 17 (1916): 101, 102 – A memorial notice for the philosopher Ernst Mach*

Today the concept of energetic 'quanta' has become what could be called the 'energeticist' equivalent of the old 'materialist' notion of particles – both united by the assumption that reality is composed of discrete entities or units rather than continuous fields.

What did Einstein have to say about this?

"All these fifty years of conscious brooding have brought me no nearer to the answer to the question, 'What are light quanta?' Nowadays every Tom, Dick and Harry thinks he knows it, but he is mistaken."

"The quanta really are a hopeless mess."

Nowhere does the *mythical and quasi-religious* nature of what is taken today as 'physics' come to expression more clearly than in the attempt to 'unify' 'Quantum Mechanics' with Einstein's theories of Relativity. Quantum Mechanics has to do with the nature of electromagnetic forces – that is to say, of light in the form of 'photons' or 'quanta' of energy. Relativity on the other hand has to do with the nature of light in relation to space and gravity. 'Gravity' however, is not just as a force wholly distinct to electromagnetism but – not least in its now most problematic form, that of so-called 'dark matter' – now appears as a 'dark' and mysterious counter-pole to light itself. Recognising this, we can begin to see that all the abstract and arcane terminologies, mathematical theories and competing theoretical frameworks of modern physics are but a modern echo of *age-old religious mythologies of a universe created from primordial forces of Light and Darkness*. Needing as it does to 'scientifically' explain their relation through a single 'Unified Field Theory' or 'Theory of Everything', the quest for such a Theory is now the most important *theological* challenge facing physics in its attempt to shed 'light' on the fundamental nature of reality – for all its attempts to do so are now threatened by the mystery of 'dark' matter.

Light has long been a primordial religious metaphor of 'spirit' – hence such phrases such as 'en-lightenment', 'illumination' etc. Only in Indian religious and philosophical thought however, arose the decisive recognition that physical light - the light of suns, stars and galaxies – would not itself be anything visible or measurable without an awareness of it. Hence arose a concept of the essence of light as *awareness itself* – not an awareness that is the mere 'property' or 'emanation' of divine beings or cosmic bodies, material particles or quanta of energy but an awareness pervading the universe *as* space and light, an awareness of which all things and all beings are ultimately composed. It is only through and within this universal awareness that all and any 'actual' phenomena can first 'come to light' as manifest phenomena – from within the 'gravitational' density and 'darkness' of a realm of infinite potentialities of actualisation and manifestation.

In that grand synthesis of Indian metaphysical and religious thought forged in 10th century Kashmir by great religious metaphysical thinkers such as Abhinavagupta and Kshemaraja we already find a more foundational and fundamental understanding of both space and light than any that can be found in modern physics – one which understands them not as dimensions of an 'objective' universe of 'matter' or 'energy', but rather as dimensions of an essentially subjective universe – of subjectivity or awareness as such in its universal, all-pervasive and 'field' character:

"The being of all things that are recognised in awareness in turn depends on awareness."

"… space is inherent in the soul as true subjectivity which is at once empty of objects and which also provides a place in which objects may be known." (Sri Abhinavagupta)

"Every appearance owes its existence to the light of awareness. Nothing can have its own being without the light of awareness."

Kshemaraja

Space as such, in other words, is nothing essentially physical but rather a universal and continuous field of *awareness* in which alone all phenomena can first 'come to light' from a 'dark' realm of unbounded potentialities *of* awareness. Light too, is nothing essentially physical but rather metaphysical – physical light being an expression of the metaphysical *light of awareness* in which all things first 'come to light', indeed in which they first come to be or emerge (to 'come to light' being the root meaning of the Greek verb *phainesthai* from which the very word 'phenomenon' derives, and 'to emerge' being the root meaning of the Greek verb *phuein* – from which the very term 'physics' itself derives).

Light, including the light of suns, stars and galaxies, is something that 'comes to light' only *in the light* of an awareness of it. The light that we see around us is in turn nothing but a manifestation of the transcendental-metaphysical light *of* that awareness.

Not only does physics, as – supposedly – the most 'fundamental' of all the sciences, fail to explain the 'hard facts' of our everyday subjective experience of phenomena such as light, it fails above all to recognise the most fundamental or 'hardest' scientific 'fact' of all which is not the 'objective' existence of a universe of matter, energy, space and time, light and gravity – but rather a *subjectively* experienced *awareness* of such a universe. In this sense, physics is no more based on hard fact than religious 'mythologies' of light and darkness.

As I have argued in my book *The Awareness Principle* the unseen and unanswered philosophical challenge to current science lies in the recognition that awareness cannot – *in principle* – be the product or property of any thing or being we are aware *of* – including our own body or our own being.

This simple logical principle means that awareness alone – and not any form of physical matter, energy, space or time, must itself be the sole and absolute reality behind all things – and thus also the sole possible basis for a 'Theory of Everything'.

A new and true concept of science must therefore be founded on a new fundamental principal – what I term 'The Awareness Principle'.

This Principle recognises (1) that *awareness is everything* and (2) that everything in turn *is* an awareness – a *manifestation* of consciousness and no mere 'object' of consciousness.

The hard fact that current physics has yet to confront is that the only possible 'Unified Field Theory' is a Unified Field Theory of *Awareness* – of 'subjectivity'. Yet acknowledging this fact challenges the most fundamental of all the unquestioned religious dogmas of modern science – indeed its 'sacred cow'. This is the dogma that 'truth' is 'objectivity' and that knowledge is by definition knowledge of 'objects' on the part of isolated 'observers' or 'subjects' – subjects who happen to have mysteriously and inexplicably 'evolved' a subjective awareness or 'consciousness' from out of an otherwise wholly insentient and *unaware* universe of 'objective' space and time, 'matter' and 'energy'.

What I call 'The Science Delusion' is the delusion of an 'objective science' which stands in the way of a new model of science – as *subjective science*. Subjective Science is also 'Qualitative Science'. I call it *Cosmic Qualia Science*. It constitutes a 'Second Scientific Revolution' – restoring science to its empirical roots in a direct subjective and *qualitative* experience of phenomena – and constituting a science of 'qualia' rather than of abstract 'quanta'.

The term 'qualia' is conventionally used only to describe our experience of the outer sensory qualities of things. Such sensory qualities however, are in essence the outward manifestation of innate 'psychical' qualities of subjectivity or awareness as such. This distinction is crucial. For whilst a feeling of physical warmth is clearly a 'physical' or sensory *quale* what we experience as 'warmth of feeling' is a psychical *quale* or 'soul quality'.

Light, from this point of view, is essentially a manifestation of the light of awareness. Similarly, outwardly perceived colours are expressions of felt 'colourations of awareness' – comparable to differently coloured subjective moods. A qualitative, subjective-scientific understanding of light and darkness offers a quite different route to the understanding of 'dark matter' and of 'gravity' itself – the latter being nothing but a density of as-yet unmanifest qualities and qualitative intensities of awareness.

Awareness, like light, is first of all an awareness of *potentiality* – of what could *potentially* be illuminated or 'come to light' as an actual phenomenon from within a 'dark' realm of potentialities *of* awareness. Yet if what physical science seeks to 'bring to light' is already pre-defined in terms of metaphors deriving from the realm of *actual* rather than potential phenomena (for example the metaphor of electromagnetic or gravitational 'waves') it will forever leave us 'in the dark' – not least in relation to 'dark matter'.

In contrast, and as argued in the first part of this essay, Subjective Science offers us an old-new concept, not just of 'dark matter' but of matter itself, understood as something essentially 'dark'. That is because the idea of material 'substance' is merely a mental construct created from our experience of potential dimensions of sensory experiencing – in particular the sensed potential to experience things in a *tactile* as well as visual way (for example as possessing qualities of hardness and softness, weight and lightness etc.). Thus the fact that the blind can sense things in space even without seeing them as visible, light-reflecting or light-radiating objects does not mean that what they are sensing is 'dark matter' in the new cosmological sense. Rather it confirms, that visibility is not the fundamental criterion defining the essence of matter, and that so-called 'dark matter', far from being a mysterious sub-species of matter belongs to the hidden essence of matter as such, an essence that will remain forever hidden – 'occult' – to a purely objectivist and 'matterphysical' approach to science, as opposed to a subjectivist and metaphysical one.

That is why the term 'dark matter' essentially names what physicists themselves recognise as a dark and 'occult' threat to the *hidden metaphysical framework* of 'physical' science as a whole – a framework of constructs and dogmatic assumptions which to which this 'science' is blind, and yet one to which it still stubbornly, blindly and religiously clings.

For as Heidegger noted: "Science *is* the new religion."

Other books by Peter Wilberg

Dreams, Music and the Many Faces of the Soul – *A Memoir of Metaphysical Experiences*

The QUALIA Revolution – *from Quantum Physics to Qualia Science* Second Edition, 2008

The Science Delusion – *Why God is Real and Science is Religious Myth* 2008

Event Horizon – *Terror, Tantra and the Ultimate Metaphysics of Awareness* 2008

The Illness is the Cure – *an introduction to Life Medicine and Life Doctoring* – *a new existential approach to illness* 2012

From PSYCHOSOMATICS to SOMA-SEMIOTICS – *Felt Sense and the Sensed Body in Medicine and Psychotherapy* 2010

Heidegger, Medicine and 'Scientific Method' – *the unheeded message of the Zollikon Seminars* 2005

Meditation and Mental Health – *an introduction to Awareness Based Cognitive Therapy* 2010

The Therapist as Listener – *Heidegger and the Missing Dimension of Counselling and Psychotherapy Training* 2005

Being and Listening – *Counselling, Psychoanalysis and the Ontology of Listening* 2013

The Awareness Principle – *a Radical New Philosophy of Life, Science and Religion* 2008

Tantra Reborn – *The Sensuality and Sexuality of our Immortal Soul Body* 2009

The New Yoga of Awareness – *Tantric Wisdom for Today's World* 2009

Heidegger, Phenomenology and Indian Thought 2008

Deep Socialism – *A New Manifesto of Marxist Ethics and Economics* 2003

From New Age to New Gnosis – *Towards a New Gnostic Spirituality* 2003

Head, Heart and Hara – *the Soul Centres of West and East* 2003

Available from:
www.amazon.co.uk and www.amazon.com

Articles by Peter Wilberg

The Language of Listening
Journal of the Society for Existential Analysis 3

Introduction to Maieutic Listening
Journal of the Society for Existential Analysis 8.1

Listening as Bodywork
Energy and Character; Journal of Biosynthesis 30/2

From Existential Psychotherapy to Existential Medicine Journal of the Society for Existential Analysis 22.2 July 2011

www.ingramcontent.com/pod-product-compliance
Lightning Source LLC
Chambersburg PA
CBHW051450170526
45166CB00001B/185